DECARBONIZING
ASIA

Innovation, Investment and Opportunities

Other Titles on Sustainability and Climate Change

Waking Up to Climate Change:
Five Dimensions of the Crisis and What We Can Do About It
by George Ropes
ISBN: 978-981-124-623-4
ISBN: 978-981-124-754-5 (pbk)

Sustainability: Business and Investment Implications
by Diane-Charlotte Simon and Alexander S Preker
ISBN: 978-981-124-091-1

Last Call: Humanity Hanging from a Cross of Iron and
Our Escape to Another Planet
by Daniel R Altschuler
ISBN: 978-981-125-361-4
ISBN: 978-981-125-438-3 (pbk)

Buying Time for Climate Action: Exploring Ways around
Stumbling Blocks
edited by Jan Wouter Vasbinder and Jonathan Y H Sim
ISBN: 978-981-124-918-1
ISBN: 978-981-124-977-8 (pbk)

Electricity: Humanity's Low-carbon Future —
Safeguarding Our Ecological Niche
by Hans B (Teddy) Püttgen and Yves Bamberger
ISBN: 978-981-122-435-5
ISBN: 978-981-122-930-5 (pbk)

Carbon Finance: A Risk Management View
by Martin Hellmich and Rüdiger Kiesel
ISBN: 978-1-80061-101-6

DECARBONIZING ASIA

Innovation, Investment and Opportunities

Edited by

Tony Á. Verb
Roman Y. Shemakov

Carbonless Asia, Singapore

Contributions by
Alexandra Tracy, **Sandro Desideri**, **Eric Chong**,
Bill Kentrup, **James Kruger**, **Moon K. Kim**,
Alberto Balbo, **Christine Loh**,
Archawat Chareonsilp, **Suede Kam**,
Rachel Fleishman, **Amarit Charoenphan**,
and **Davide A. Nicolini**

World Scientific

NEW JERSEY · LONDON · SINGAPORE · BEIJING · SHANGHAI · HONG KONG · TAIPEI · CHENNAI · TOKYO

Published by

World Scientific Publishing Co. Pte. Ltd.
5 Toh Tuck Link, Singapore 596224
USA office: 27 Warren Street, Suite 401-402, Hackensack, NJ 07601
UK office: 57 Shelton Street, Covent Garden, London WC2H 9HE

British Library Cataloguing-in-Publication Data
A catalogue record for this book is available from the British Library.

Illustrations by Dániel Korcsmár

DECARBONIZING ASIA
Innovation, Investment and Opportunities

ISBN 978-981-126-386-6 (hardcover)
ISBN 978-981-126-466-5 (paperback)
ISBN 978-981-126-387-3 (ebook for institutions)
ISBN 978-981-126-388-0 (ebook for individuals)

For any available supplementary material, please visit
https://www.worldscientific.com/worldscibooks/10.1142/13070#t=suppl

Desk Editors: Gregory Lee/Amanda Yun

Typeset by Stallion Press
Email: enquiries@stallionpress.com

To Paul.
To our loved ones.
To our friends in Ukraine and Hungary.
To everyone who is ready to commit energy
and resources to decarbonization.

ABOUT CARBONLESS ASIA

Carbonless Asia is an innovation and investment platform whose mission is to channel as much capital as possible into scalable technology businesses that have a strong carbon angle and measurable carbon emissions reduction potential. The Singapore-headquartered firm works with strategic corporate partners across the region with a singular focus on decarbonization, innovation, and investment opportunities. Its establishment in 2020 was inspired by the fact that there was no such specialist in Asia-Pacific. Carbonless Asia also runs thought leadership initiatives to bring leaders and practitioners together for free flow exchange and to share best practices.

FOREWORD

The last couple of decades have transformed Asia-Pacific more than any other part of the world. Some countries emerged and improved standards of living in the wake of hard work and brilliance, some others in the wake of exploration and exploitation of natural resources from palm oil to crude oil.

More recently and more quietly, East Asia also embarked on a green industrial revolution. Solar has penetrated practically all markets, with China leading both production and capacity installation globally. Hydropower is also on the rise, with Vietnam leading the charge. There is significant investment into green hydrogen capacities from Australia to Japan, but also in the Middle East.

Both the public and the private sector stakeholders know how high the stakes are. The current epoch is going to create green industrial and economic winners and losers, both amongst countries and market competitors. Much of that development and competition is underpinned by the megatrend of decarbonization. There is little to wonder about the net zero commitments and new projects popping up in the wide region every other week, whether it be a new infrastructure project or a climate-positive investment fund.

For international business people residing in the East — like myself — the speed of the change is not surprising and the global significance is obvious. What is surprising though is how little the rest of the world is taking notice of Asia. There is more news in the international media about the pollution challenges in China than about its global leadership positions in electric vehicles, solar, and wind.

These trends must be shared and must be talked about. Decarbonizing Asia is one of the most obvious and most exciting investment and development opportunities of the 21st century. The risks are rather low and the incentives are incredibly high, both in financial and planetary terms.

As more than 50% of the world's CO2 emissions are from the Asia-Pacific region, the scale of the problem and the necessary solutions are unmatched globally, along with the negative externalities if nothing actually happens. If there is no sufficient action and investment into decarbonizing the region, the global efforts to tackle climate change are doomed. On the contrary, making the right choices and decisions in and for Asia, country and industry leaders can emerge as global transformers and literal heroes.

All of the above make this book so timely and so important. It sets the scene for the massive decarbonization transition that is about to unfold in the region. It identifies and celebrates players that lead by action. It also focuses on what really matters: opportunities in investment and in innovation. Because in the end, beyond clear and courageous policy and business leadership decisions, all we need for decarbonization is more capital and more deployment of the latest best practices and innovative solutions.

I know some of the authors personally. They are credible professionals in their industries and great people in life. I wholeheartedly recommend this book for everyone who is hungry to learn about and ready to be part of the opportunities of decarbonizing Asia. The 21st century belongs to you.

John Walker AM
Chairman, Eastpoint Partners
Former Chairman, Macquarie Capital Asia

ABOUT THE EDITORS

Tony Á. Verb is a Hungarian entrepreneur and investor based in Hong Kong and Singapore with more than 10 years of experience in Asia. He is the co-founder of Carbonless Asia, an innovation and investment platform whose mission is to help decarbonize the region's cities and industries with the largest carbon footprint. Tony is also a partner of GreaterBayX, a China-focused urban technology investment platform. He holds a Master's degree in International Relations from the Corvinus University of Budapest, with studies in Italy and Singapore on scholarship as well. Amongst other accolades, he is a proud alumni member of the Global Shapers Community of the World Economic Forum. In his spare time, Tony contributes to philanthropic projects and to various publications as a writer. He is a media producer and a documentary filmmaker.

Roman Y. Shemakov's life work focuses on the intersection of jurisprudence, planetary-scale computation, and urban geotechnology. A high honors graduate in Economics and History from Swarthmore College, US, he spent a year in Taipei as a Henry Luce Scholar. Subsequently, he worked under a grant from the Harvard-Yenching Center and Carnegie, Ford, and Rockefeller foundations, investigating urban management platforms at Zhejiang University, China. He has conducted research at the Strelka Institute, Arcosanti, and Earthship Biotecture. In 2021, Roman co-authored *The Digital Transformation of Property in Greater China* (World Scientific Publishing). He is currently covering the war in Ukraine for *Global Voices* and is pursuing a Master of Law at the Yenching Academy of Peking, China.

ABOUT THE CONTRIBUTORS

Alexandra Tracy is President of Hoi Ping Ventures, Hong Kong, which she founded to provide research and consultancy on green finance and sustainable investment in Asian emerging markets. She is currently Private Sector Observer to the World Bank's Climate Investment Funds, having served as Active Private Sector Observer to the United Nations Green Climate Fund for several years. She is also Chairman of the Financial Services Research Group, an independent think tank in Hong Kong. Alexandra is a director of RIMM Sustainability Pte Ltd in Singapore and sits on a number of advisory boards, including Carbonless Asia.

Sandro Desideri is a former GE Energy executive with more than 30 years of experience in the sectors of power generation, oil and gas, smart grid, clean tech, renewables, and smart city. He has 28 years of experience in doing business in Asia. Sandro is co-CEO of Ianus Energy Transition (IET), an Asia-based firm focused on energy transition and net-zero projects, and co-founder of Link4Success Ltd, a Hong Kong firm that supports the scale up and commercialization requirements for original equipment manufacturers, small and medium enterprises, and startups operating in the sector of technological innovation. Sandro is also a senior advisor of Carbonless Asia.

Dr. Eric Chong served as President and Chief Executive Officer (CEO) of Siemens Ltd in Hong Kong from 2011 till his retirement in 2020. Dr. Chong has extensive experience in the Asia-Pacific region, gained from his over 30-year career in the infrastructure, energy, and healthcare sectors. Dr. Chong is currently an advisor to Carbonless Asia. Prior to his retirement, he served in various civic and industry bodies in Hong Kong, including the Business Environment Council (Deputy Chair), Climate Change Business Forum Advisory Group (Chair), Construction Industry Council's Technology Center, Cyberport Advisory Panel, and HK Trade Development Council.

Bill Kentrup has been involved in Asia's environmental and infrastructure markets since 2001, first as a founding director of Noble Environmental Solutions and Carbon Credits, and later as Head of Asia Environmental Financial Products at Macquarie Bank and Head of Asia Renewables at Macquarie Capital. Bill has cultivated a broad and trusted network of sustainability leaders in capital markets, the private sector, and government. More recently, he has partnered with leaders in the blockchain space and is now focusing on applying blockchain technology to the financing of infrastructure and environmental assets through the establishment of Allinfra.

James Kruger is a former investment banker, global general counsel, and integrity officer for Macquarie Group Limited. He now consults for companies in the supply chain for critical minerals, helping with the financing of new projects and the industry and policy settings, so to expand and develop new ESG-effective supply chains. He is also a technology investor and is the chairman of ASX-listed investment vehicle Powerhouse

Ventures Limited, as well as a board member and advisor for several deep tech companies that help solve global problems.

Moon K. Kim is a Seoul-born career sustainability specialist with decades of experience in the energy and environmental sectors, first as an environmental engineer, then as a strategic consultant, entrepreneur, investment banker, and venture capital/growth equity investor. She has been involved in more than USD 1 billion of transactions around the world, supporting growth-stage companies committed to climate change mitigation and adaptation. Moon holds an MBA in Finance from The Wharton School of the University of Pennsylvania and a BS in Environmental Engineering from Harvard University.

Alberto Balbo is co-founder and co-CEO of Ianus Energy Transition (IET), an Asia-based firm focused on energy transition and net-zero projects. He previously held international business executive roles with multi-disciplined, Fortune 500 conglomerates such as General Electric and Honeywell. Alberto has robust international experience across industries such as power generation, oil and gas, mining, manufacturing, digital, and software and services. Alberto holds degrees in engineering and business, is a member of the Australian Institute of Company Directors and a non-executive director of the Australian Microgrid Centre Of Excellence (AMCOE).

Dr. Christine Loh is the chief development strategist at the Institute for the Environment, Hong Kong University of Science and Technology. A lawyer by training and commodities trader by profession, Dr. Loh has spent the bulk of her career in public policy and politics, first serving as a legislator and then as a government minister in the Environment Bureau in Hong Kong. She was the CEO of a policy think tank and is a published author of many academic and popular works. She is a director of for-profit and non-profit organizations, and she also advises multinational companies.

Archawat Chareonsilp is a veteran Corporate Executive, avid climate change student, and serial entrepreneur who invests in and builds regenerative technologies in energy, agriculture, and real estate development. He formerly held regional senior leadership roles in sustainability at The Coca-Cola Company and Pepsico. Archawat believes that while climate change may pose serious challenges, it also ushers in a new way of thinking, designing, and living, which offers new growth opportunities and the sustainability of future generations.

Suede Kam helps innovative organizations such as the United Nations, Salesforce, and Huawei bring emerging technologies and special projects to new markets. She is the founder of Pine Lea Limited in Hong Kong, which connects tech brands global to Asia, and provides go-to-market strategy and operation consulting. Suede also advises some of the world's fastest-growing brands and impact-positive startups for scale and strategic partnerships. She strongly believes in equitable access to information and also supports change-makers in their sustainability efforts. Suede is an advisor to Carbonless Asia.

Rachel Fleishman advises corporate, nonprofit and government entities on sustainability, climate change and climate security. She currently works with sustainability intelligence platform HowGood and the Center for Climate and Security in the US. Over her 14 years in Hong Kong she helped launch Carbonless Asia; spearheaded sustainability for BASF Asia-Pacific; and directed the Climate Change Business Forum. Rachel holds a BA from

Tufts University, an MA from the School of Public Policy, University of Maryland, and an MBA from the Kellogg School of Management, Northwestern University.

Amarit Charoenphan is the former Chief Transformation Officer at Enserv Holding, Thailand, an Energy Solutions Provider whose core businesses lie in clean energy innovation and generation and non-energy innovation encompassing health and living solutions. He is also the Head Coach and Partner at Omelet (an award-winning innovation builder based in Bangkok) and ASEAN Director of Impact Collective, a community-driven investment and acceleration project for impact-driven startups focusing on opportunities in Asia.

Davide A. Nicolini spent several years in the space transportation sector on electric ion engines and advanced spacecraft development, space rocket development, and spaceport operations. He has extensive experience in large space infrastructure programs — he oversaw development, operations, and value chain enhancement, and led innovation in the space industry, especially in areas concerning either the increase of efficiency via machine enhancements or the reduction of carbon footprint of manufacturing and spaceport processes.

ACKNOWLEDGMENTS

This book is an important milestone in my personal and professional development, bringing me to an important junction in life. I would like to highlight and thank:

Paul Schulte, for introducing Roman, and Roman for being a great wingman.

Bill Kentrup, for inviting me to the land of decarbonization, and all Carbonless advisors, along with all contributors of this book, for teaching me new depths and breadths of it: Aim, Alberto, Alex, Barak, Christine, Cristina, Davide, Eric, James, Moon, Nazia, Ohm, Pat, Rachel, Sandro, Suede, and Swamy.

John Walker, for leading by example and for the special support.

Dani, for bringing his art and talent to this book.

James Gifford, for the original spark and ongoing friendship.

The Felkai family and Felkai Judit specifically, as well as Veres Erzsébet — without your intervention, I might not find myself where I am today.

My family for their unconditional love and support in everything I do.

Natalie, for being my north star, for the patience, and for teaching me patience.

Tony Á. Verb

I want to thank the following people who taught me incredibly valuable life lessons this past year. It is likely I will have to relearn them constantly.

Paul Schulte, for the wisdom and reflection.

Ling Li, for standing by one's principles.

Nicolay Boyadjiev, for building something meaningful.

Tony Verb, for the vision and patience.

Filip Noubel, for an unequivocal love for the world.

Those who made this work possible through constant inspiration, contemplation, and occasional brachiation: Flora, Virginia, Sean, Stefan, Felix, Kola, Max, Danya, Alexandra, Uliana, *et al.*

As always, thanks to Ben Shindel, Jonathan Kay, Emma Hobbs, David Pipkin, Halo Lahnert, and Benjamin Jackson for the always-appreciated mental gymnastics.

Thank you to Paul Hsu and Wei Feng Lee for a valuable life lesson.

Particular thanks to Branden Yeats for his compassion for the cause of sovereignty and regular passion for niche history.

To Neal, who has danced with life.
To Olga, who will be a wonderful mother.
To Olena, who already is one.

Roman Y. Shemakov

CONTENTS

PROLOGUE

by Tony Á. Verb

We are Not here to Talk about Climate Change

If you are reading this book, you have most likely heard about climate change and you acknowledge that action is needed, and it is needed urgently. There is overwhelming scientific support, multinational commitments, and probably thousands of books to prove that point, so we will not waste a single paragraph on that ourselves. Thankfully, even the mainstream media has caught up with science recently and less and less people need convincing about the facts: climate change is happening. This book comes at a time when pragmatic, practical, and scalable climate solutions need a push in the right direction. Asia's carbonless century will most likely be heralded by technology startups, right incentives, groundbreaking inventions, and new best practices.

This is a good place to be. Over the last couple of decades, since the first green activist movements and the first Earth Day in 1970 to the very recent Extinction Rebellion movement, the challenge has been to simply cut through the clutter. These days there is also a consensus around the fact that climate change is mainly caused by humans and it is not very good for us, to say the least. The very recent IPCC report[1] stated irrevocable evidence on climate change and the human factor behind it.

Thankfully, millions of people across generations and the world have broken through the barriers of public opinion and political inertia to successfully raise awareness about climate disasters. Continuous scientific support, studies, technological breakthroughs, and commercial proofs of success have helped accelerate the turning of the tides. And the news is now distributed both by mainstream and, increasingly, social media platforms, reaching and mobilizing more and more people.

With mainstream awareness rising and public opinion changing, policy has been shifting gears as well. By 2021, with the change of the United States (US) Administration, there is no G7 economy left that would not take the issue seriously, now adhering to the Paris Climate Accords. Arguably, the European Union has been leading the way in aggressive policies that set the bar for the rest of the world and we have arrived at the point where countries are trying to outbid each other in their commitments to fight climate change. We shall hope it is serious and not simply showmanship.

Where public opinion meets policy change, business follows. Some industries are answering increasing consumer demands to shift to more climate conscious operations and products. In other sectors, policy change or the prospect of it puts pressure on corporations of all types to proactively change their ways to avoid risks. Capital markets reward such behavior either via more favorable green instruments or simply via private equity-type bets on companies that adapt to the new realities.

This feels like the perfect storm and there is a lot to be optimistic about. The writers of this book certainly do not feel the need for further fearmongering; there has been enough of that. We are committed to focusing on the opportunities and the solutions that will get us to net zero, or as close to it as possible. We do not mind the naysayers who say that it is too late and impossible. The same IPCC report that stated that humans are the reason for the undeniable climate crisis also states that there is evidence that it is not late to act and it is possible to turn things around. A recent IEA[2] report suggests that more than 80% of the technology needed to reach 2030 targets on the path to net zero is already available, and the rest is already in development. We need to believe, and we have all the reasons to be certain that the change is possible. What we need is focus, commitment, and scalable solutions.

In this book we direct all attention to one metric: CO2. It is widely accepted that decarbonization is the most critical process we must achieve on a global scale to tackle climate change. Methane and all other greenhouse gases (GHGs) are important, but one must focus, and so we picked the biggest culprit. With regard to commitment, we will highlight investors, governments and serious industry players, i.e., structural emitters, that are allocating resources to change their ways, lead by example, and profit from the changing landscape. And when it comes to scalable solutions, we will be highlighting startups, technology companies, inventions, and best practices that are ready to be implemented across borders.

We do this with a focus on Asia-Pacific. Whilst more than half the planet's CO2 emissions are from the region, with the single biggest emitter being China, there has been very little dialogue and information about the region's past, present, and future in the context of decarbonization, innovation, and investment opportunities. This is all the more surprising if one considers that China has played a significant role in bringing down the cost of photovoltaic systems to the point that renewables are now the cheapest source of energy if deployed properly. The country leads the world in installed solar and wind capacity as well. Elsewhere in Asia, a Singaporean company is building the world's largest floating solar power plant in Indonesia, whilst Australia is investing heavily to become a hydrogen superpower. Out of the top 10 countries with the most installed renewable energy capacity in the world, six are in Asia-Pacific. We believe the lion's share of effort in decarbonization must come from the region, and this process will lead to systems, industries, and lifestyles that will define the second half of the 21st century and perhaps even beyond.

We are Here to Talk about Opportunities

About This Book

Hence, we decided to write this book to be the first one that collects the developments, facts, innovators, and opportunities in Asia-Pacific in the context of decarbonization. We also scheduled this work around COP26, the climate conference that virtually everyone in global politics, sustainability, and asset management has awaited with sky-high expectations. There was a buzz in the air and a shared feeling that 2021 and the event itself would prove to be a significant milestone in combating climate change and on the course of decarbonization. This hope contrasted with the lengthening despair over COVID-19 and recent geopolitical turbulence and was a very much-needed source of international collaboration. People had something to look forward to that could propel positive action on a global scale. Some of those expectations towards the event have been fulfilled and outdone; some have not.

The biggest development from our perspective was that COP26 was probably the first climate conference that received mainstream media attention before it even happened and became a talking point in everyday conversation. A lot of people cared about the outcome of this event and when expectations are high, disappointments happen. No one can ignore though, that India, Saudi Arabia, Thailand, the United Arab Emirates, and Vietnam have all timed the announcement of their net zero commitments and targets for and around the conference. For many participants, observers, and stakeholders, there was a lot more to be hopeful about before and after the conference.

More than a dozen of those, not only hopeful but outright optimistic people joined forces to work on this book. Our goal has been to create an overview and give you a snapshot of various industries and how they are expected to change over the coming decades in the net zero context. We are academics, engineers, investors, and former policymakers with deep and diverse experience in the Asia-Pacific region. We have a shared interest in decarbonization through technology and innovation. We also share a passion to create value at this intersection now and keep sustaining it for years to come. We want to contribute to the fight against climate change, but with a positive mindset, looking at the opportunities. We strongly believe that decarbonization is not only one of the biggest challenges of this decade but also a trend that will reward entrepreneurs, investors, and anyone with resources to allocate.

We wanted to be comprehensive with this book and explore the sectors with the biggest carbon footprint and potential impact on decarbonization. To make this an enjoyable read for everyone, we tried to stay away from technical jargon, even when it proved to be difficult, and stuck to a "less is more" principle. We believe that the biggest value of this work is to collect trends, facts, and expectations of various interconnected sectors and shine the light on some of their key players in the region in a way that is digestible and engaging for you, whether you happen to be a university student, an investor or a policymaker.

Indeed, throughout the book, we have remained true to our multi-stakeholder approach, talking to and highlighting corporates, governments, startups, and investors to give a balanced and full picture. It has been an enlightening and educational journey for us all, and it is my duty in this paragraph to recognize every single Contributor's effort, intelligence, and collaborative spirit that shone through the months of hard but truly enjoyable work. On behalf of us all, I thank you.

The Outline

As our ultimate focus is maximizing impact, we decided to look at the various factors and actors that can contribute to decarbonization the most.

Humans

This first section — just like all the others — is divided into three chapters, with each chapter covering one broad topic, typically around one industry. Our first section is dedicated to us, humans. We have been creating this wave of climate change, and we can and must stop it. The industries that are the biggest emitters are designed to benefit us (or some of us), and the consumption that is driving those industries is done by us. Similarly, the policies we create, the investment decisions we make, and the coordinate system in which we make those decisions are all up to us. If those rules are not made in a carbon-conscious manner, there will be more emissions and worse climate impact. Otherwise, it will be the opposite, and all the opportunities will be within a carbon-conscious rule system.

That is why we start off by exploring the decarbonization-related trends in policy and finance in Asia-Pacific. We believe they are the single most important factors in our discourse. It will be followed by "Lifestyle & Consumption", an area that has received a lot of positive attention lately with the hot investment themes of alt-proteins and the zero-waste movement. Next, we will look behind the products we consume and get more technical by going deep into the dirty topics of "Industry & Waste".

Cities

When we discuss anthropogenic — human-activity induced — climate change, the urban context is an important one. Although only occupying 3% of the planet's land mass, cities contribute to 60% of the GDP output and 70% of global CO2 emissions.[3] As Asia-Pacific is rapidly urbanizing with new cities being built and old ones rebuilt and expanded, it is critical that the process happens along with sustainable principles, applying low carbon solutions. As best practices can bring long-term benefits, bad practices get embedded in urban structures for decades, and this is something we must avoid.

In the second section, we cover "Energy", "Mobility", and "Built Environment" as they underpin urbanization and, collectively, these sectors represent more than 70% of the planet's CO2 emissions. Given the associated carbon footprint and the potential positive spillover effects and upside, we have a real chance of hitting the global decarbonization targets if we get cities and the related industries right over the coming years and decades.

Nature

Ultimately, what anthropogenic climate change is impacting negatively is the composition of our atmosphere, nature, and the entire balance of the planetary ecosystem. This is one reason why we must dedicate attention to nature when we discuss decarbonization. Another reason is that nature-based solutions are crucial in scaled decarbonization efforts. With the right measures in place and best practices adopted on scale, oceans, forests, and soil alone can sustainably absorb more than 60% of the current CO2 emitted globally. Protecting them and adding innovation where possible and necessary is critical to meeting decarbonization targets in the Asia-Pacific region as much as anywhere else around the world.

In this section, the three chapters cover "Forestry & Oceans", "Sequestration", and "Food & Agriculture". From ocean sinks to geoengineering and direct air capture to carbon mineralization, there are a myriad of opportunities to take carbon out of the air and store it safely, permanently, efficiently, and in a useful manner. Some solutions are more scientific and riskier, others are industrial and expensive, but there are solutions nonetheless that are out there and ready to scale.

The Structure

You will notice that we follow the same logic across the sections, broken down into three chapters each, with each chapter having four bite-sized content pieces. The core of each topic is explored in an analysis piece, which sets the context through foundational definitions and by exploring the historical achievements and facts for any given industry. We then provide a snapshot of present trends and updates, finally turning our attention to what we really wish to focus on: the future and the opportunities.

Each chapter comes with some color and specific best practices in the form of two case studies, one dedicated to a startup, another to a large corporate player in the industry. We also interview at least one decision-maker, thought leader, or researcher for each topic, in order to expose multiple perspectives along the way. We shall start this book with a warm-up exercise. To show you how this structure works, we will set the core definitions for the entire book and summarize the past, present, and future of decarbonization and technology in Asia-Pacific.

Decarbonization in Asia-Pacific

Definition

As we discuss decarbonization in Asia-Pacific throughout the various chapters, we must make sure that we speak the same language. Let us set the core terms herewith.

Asia-Pacific

Our geographical definition of Asia-Pacific is as broad as it gets, including the Middle East, South Asia, Central Asia, all of East Asia, as well as Australia and Oceania, but not Russia.

Decarbonization

Either reducing or removing CO_2 from the economy and its various sectors specifically. This book believes decarbonization is desirable and its Contributors are aligned on the goal to positively influence the process.

Innovation

In this book beyond technology and digital solutions, we are open-minded to entertain various forms of innovation with a decarbonization impact: new materials, new business models, and also smart policies.

Net Zero vs. Carbon Neutral

Both colloquially and in journalism, these two terms tend to be used inter-changeably, but there is a difference. Whilst Carbon Neutral means that an entity removes an equal amount of CO_2 that it emits; in the case of Net Zero, the requirement covers all other GHG emissions as well.

Carbon Negative vs. Climate Positive

The two terms have the same meaning and refer to any entity going beyond achieving carbon neutrality, removing further CO_2 from the atmosphere, creating a negative carbon balance.

Past

The Asia-Pacific region has been rather notorious for its pollution-prone economies — it is not only China, where soil contamination and toxic air in cities since the advent of its economy's boom have been very well documented. According to a number of studies,[4] the vast majority of the cities with the worst air quality record are from our region. That is true in Ulaanbaatar, Mongolia's capital, and in Delhi, which has ranked No. 1 on the global rankings of the worst air pollution levels amongst capital cities for years. The challenges are multifold, as in Australia, the mining industry has been a significant pollutant, whilst deforestation has been a major issue all across Southeast Asia, where Malaysia alone has lost a fifth of its forests since 2001.[5] Creating a complete list of issues — and issues that have barely been addressed — in the region that one way or another relate to climate change and sustainability is beyond the capacity of this book.

When it comes to carbon footprint specifically, eight out of the top 10 countries, according to their CO_2 emissions per capita, are in the region. In that regard, Qatar has been consistently leading the global ranking for decades, most recently with 32.4 tons annually (pa), more than seven times over the global average. Unsurprisingly, states with heavy reliance in their economies on fossil fuels, like Bahrein, Brunei, Saudi Arabia, and the United Arab Emirates (UAE), are similarly high on the list.

Over the last couple of decades, the region has been busy industrializing and catching up on economic and well-being terms with the developed world, typically relying on industries with high carbon footprint. The question of climate change, pollution, and decarbonization, in particular, has not been at the top of the agenda. That has started to change recently.

Present

Whilst China is now the second-biggest economy in the world, the majority of the countries in the region are classified as developing economies by the UN,[6] with little representation in the top echelons of global rankings and forums. For instance, Japan is the only country representing Asia-Pacific within the G7, and only Hong Kong SAR and Australia tend to make the top 10 in the global HDI rankings, with the 4th and 8th positions, respectively,[7] in the most recent UNDP study. Two of the twenty worst performers, Afghanistan and Yemen, are also in the region, which shows the diverse levels of development and thus challenges and priorities across different borders, markets, and cultures. As such, when we discuss decarbonization in Asia-Pacific and when one is tempted to draw conclusions on the region's performance and commitment to achieving carbon neutrality, this diversity and the different priorities must be taken into consideration.

With that in mind, it is even more impressive when China (the planet's biggest emitter with 28% of the global share[8]), Hong Kong, Japan, and South Korea all moved to make their carbon neutrality commitments in 2020 pretty much around the same time. This was a dramatic change not only because these economies represent more than one-third of global emissions, but because they signaled a new era of economic development to everyone else. Decarbonization-related industries, their vast array of new products and services from electric vehicles (EVs), through alternative protein to green finance are the new battlegrounds in international competition. These are also the areas where economists and politicians may expect growth and job creation from. Decarbonization has become an economic and political priority across the region. New policies, new plans, and new investments are revealed non-stop; and this is a great thing.

Some countries with no specific carbon neutral commitments also seriously develop and invest in future proof, sustainable industries — Australia is focusing on sustainable energy, New Zealand on sustainable food and agriculture, and Singapore has not only positioned itself to become the sustainable technology hub in the region but also committed to the Singapore Green Plan, encompassing almost all facets of the economy.[9]

Even in the Persian Gulf, where the Gulf Cooperation Council (GCC) member states alone have way more — 3,200 GT of CO_2 equivalent — in fossil fuel reserves than what the 1.5 and 2°C warming targets could budget for, governments have been investing in and announcing various sustainable and carbon-neutral development concepts.[10]

The UAE is committed to its Energy Strategy 2050, aiming to increase the share of clean energy to 50% by 2050, achieving it through a mixture of nuclear, renewables, and carbon capture and storage (CCS) technology applied to their fossil fuels industry. Qatar has pledged to make the 2022 FIFA World Cup, which it is hosting, a carbon-neutral tournament through an array of innovations. The UAE's Emirates Global Aluminum (EGA), a significant global aluminum producer, claimed in early 2021 to be the first in the world to produce and sell green aluminum (for BMW) by using solar power.[11] And there are the sustainable city concepts that have been built or are under development, from Masdar in Abu Dhabi to Neom in Saudi Arabia. The pressure to diversify these economies and the availability of resources — capital, sunshine, infrastructure, human capital — will undoubtedly drive new sustainable developments and carbon-neutral agendas in the region further in the coming decades.

It will be interesting how other Asian nations in the region adapt to the new carbonless world. Given that India represents 7% of the global CO_2 emissions with a high level of projected industrialization and growth, it is at the top of the watch list. Neighboring Bhutan, in the meantime, has achieved not only carbon neutrality but claimed to be the first carbon-negative country on the planet, with a commitment to keep it that way forever. This country with less than one million inhabitants is somewhat an anomaly but a good example where great carbon results do not necessarily need high technology. In this case, the low level of industrialization and the small population compared to the available land area, along with favorable conditions for a hydro-power-based renewable energy system make achieving carbon neutrality much less daunting. The country is more than 80% covered by forests and, by law, the figure may not go below 60%.[12] The awareness and commitment on a policy level is the notable factor in this case and something any government can draw inspiration from in the region and around the world.

Future

With decarbonization becoming one of the hottest words in politics and public discourse, businesses have recently started to invest heavily in the space substantially, whether to truly change their operations with conviction or for PR purposes to attract attention. The list of newsworthy initiatives is growing longer by the day. Standard Chartered, a Fortune 100 bank with a heavy presence in the region, announced in mid-2021 its intention to open its first paperless branch in Hong Kong with an emphasis on green offerings.[13] Sinopec, China's state-owned oil and gas major that is 5th on the Fortune Global 500 list,[14] went beyond announcing its own 2050 carbon neutral target by initiating a flurry of investments into solar and hydrogen infrastructure developments.[15] The company has an obviously significant carbon footprint; thus, it has a significant role to play in achieving China's carbon-neutral targets. Even more explicit is another 2021 announcement from Temasek, Singapore's sovereign wealth fund, to launch a "Decarbonization Investment Partnership" with the world's largest asset manager, Blackrock.[16] Another big global name, Bain & Co., also just launched its global sustainability innovation center in the city-state.[17]

These are only a very few examples of the recent dramatic shift in the attention on sustainability and decarbonization-related opportunities in the region by governments and local and international corporations. These very commitments and investments set the scene for the future. The aforementioned Temasek–Blackrock investment platform is targeting more than USD 1 billion assets under management, and that figure does not even seem big, considering the needs and prospects.

Bain & Co. outlined the green economy as an opportunity worth more than USD 1 trillion for Southeast Asia alone by 2030. Carbontech (i.e., technology solutions that specifically remove CO2 from the atmosphere and/or use CO2 as a raw material) is predicted to be an industry worth more than USD 1 trillion per year as well. The UN IPCC estimates that the required investment into energy systems in the context of climate change mitigation is over USD 2.4 trillion annually until 2035.[18] With all this in mind, the increased commitments and investments seem like a drop in the ocean. This is clearly just the beginning. It has to be.

The region is not all about demand, however, as there is an increasing number of global leaders emerging across sectors. Considering cities and energy transition, Shenzhen in Southern China has been set as an example to follow — noted all around the world — after it fully electrified its mass transportation system, from cabs to buses. The hardware for these fleets has been provided by local champion BYD, which gained fame in 2008 when Warren Buffet's Berkshire Hathaway took a then-surprising equity stake in the little-known Chinese battery and EV maker company. Now it is second only to Tesla in unit sales globally,[19] with its taxi and bus lines giving it a significant competitive edge and room for future growth.

Speaking of EVs, whilst not impressive on a proportional basis (3.3 per 100 vehicles on the road), China is home to almost half the planet's — "highway legal plug-in" — EVs, with over 4.5 million units sold to date.[20] Its state-owned car companies like BAIC or SAIC are shifting to EV lines, and globally recognized and US-listed startups like NIO, Xpeng, or Li Auto are making waves, projected to each churn out more than 100,000 fancy cars annually by 2025. Japanese carmakers Nissan and Toyota have been pioneers in electric and hybrid cars, with their South Korean rivals doubling down on the trend as well.

The sustainability push might not stop at transport, but perhaps start with energy. China's role is expected to be as significant in making green hydrogen (hydrogen created by using renewable energy) commercially viable as it proved to be the case with the solar industry. Bringing down the cost of electrolyzers through China's manufacturing efficiency and economy of scale, as well as by driving demand, will be crucial for hydrogen to fulfill its potentials in decarbonization; more on that later in this book. Whilst China is taking green hydrogen seriously, it is not alone. Japan recently announced that it aims to become a hydrogen superpower. Japanese firms like Kawasaki Heavy Industries and Mitsubishi have been heavily investing in related projects in Australia,[21] a country that itself is dubbed to be an ideal place for green hydrogen production,[22] which has also recently announced multiple projects to facilitate that reality.[23]

Three of the top five hydrogen exporters and importers are both in the region, and the oil and gas economies in the Middle East are investing heavily in green hydrogen production, both in words and in hard currency. Qatar opened the wave, and now the UAE and Saudi Arabia are outbidding each other for the largest related development projects.[24] Uninterrupted sunshine (i.e., abundant renewable capacity, crucial for competitive green hydrogen production) and a sophisticated energy market expertise with global clients are all in favor of the GCC countries to be global leaders in the field.[25]

With capable and well-resourced governments driving change and big corporations following as industries transform and risk-tolerant capital present, related startup companies have been popping up across sectors. And there is more than EV-makers in China. Impossible Foods and Beyond Meat, two alternate protein companies from the US, have been loud venture capitalist (VC) and public market success stories in recent years, and for a while, there were no such startups in Asia. OmniPork, a Hong Kong-based startup, however, is now expanding in the US after tremendous growth across Asia-Pacific since its launch in 2018. Its success has inspired dozens of startups across the region to launch all sorts of products, from plant-based fish to insect protein products. The fact that a restaurant in Singapore became the first in the world in 2020 to serve lab-grown chicken meat from US alternative protein startup Eat Just[26] suggests that the region's role to drive innovation in food, lifestyle, and consumption is not only embraced locally, but recognized globally.

Why This, Why Now

We could continue the list of reasons why Asia-Pacific matters in the context of decarbonization, and we could list all day the investment statistics and related innovation applied by global MNCs or developed by local startups. We shall do this throughout this book because there is a lot to tell, and somehow these stories have not been collected and shared in one book so far.

There is no wonder. There has not been much to write about. Apart from a few exceptions, the region has been known for its singular focus on growth with little regard for the environment, dominated by heavy or just carbon-heavy industries and non-sustainable practices. Until recently, the startup ecosystems and venture capital scene in Asia-Pacific — vital to technology innovation — were also rather immature. All that gradually started to change and heat up with China's lead since the early 2010s.

We are *just* experiencing a shift and are at an inflection point, where a book on decarbonization, innovation, and investments is extremely timely. This change is the result of more and more countries reaching a comfortable level of development and that are now aiming for globally recognized leadership positions and changing their economic development priorities. They are also trying to change their global image and embrace the responsibilities that their global footprint and influence entail.

It is about time. It is the Asian Century after all.

Endnotes

1. IPCC, "Climate change 2021: The physical science basis. Contribution of Working Group I to the Sixth Assessment Report of the Intergovernmental Panel on Climate Change", Cambridge University Press, Cambridge, United Kingdom and New York. doi:10.1017/9781009157896.
2. International Energy Agency, "Net zero by 2050 — A roadmap for the global energy sector", 17 May 2021.
3. Figures published by the United Nations, see https://www.un.org/sustainable development/cities/.
4. IQ Air, "World air quality report", 2020, https://www.iqair.com/world-most-polluted-cities/world-air-quality-report-2020-en.pdf.
5. Thomson Reuters Foundation, "Global rainforest loss 'relentless' in 2020, but Southeast Asia offers hope", *Eco Business*, 1 April 2021, https://www.eco-business.com/news/global-rainforest-loss-relentless-in-2020-but-southeast-asia-offers-hope/.
6. United Nations, "Country classification: Data sources, country classifications and aggregation methodology", World Economic Situation and Prospects 2014, https://www.un.org/en/development/desa/policy/wesp/wesp_current/2014wesp_country_classification.pdf.
7. United Nations, "Human development report 2020", http://hdr.undp.org/sites/default/files/hdr2020.pdf#page=357.
8. Union of Concerned Scientists, "Each country's share of CO2 emissions", https://www.ucsusa.org/resources/each-countrys-share-co2-emissions.
9. SG Green Plan, https://www.greenplan.gov.sg/.
10. Global CCS Institute, "Momentum for CCS across the Gulf Cooperation Council is growing", https://www.globalccsinstitute.com/news-media/insights/momentum-for-ccs-across-the-gulf-cooperation-council-is-growing/.
11. Bruce Stanley, "BMW to use world's first solar-made aluminum", *Bloomberg*, 2 February 2021, https://www.bloomberg.com/news/articles/2021-02-02/metal-giant-to-supply-bmw-with-world-s-first-solar-made-aluminum.
12. Climate Council, "Bhutan is the world's only carbon negative country, so how did they do it?", 2 April 2017, https://www.climatecouncil.org.au/bhutan-is-the-world-s-only-carbon-negative-country-so-how-did-they-do-it/.
13. Enoch Yiu, "Standard Chartered to open paperless 'green branch' in Hong Kong offering sustainable finance products, says local boss", *South China Morning Post*, 9 August 2021, https://www.scmp.com/business/banking-finance/article/3144276/standard-chartered-open-paperless-green-branch-hong-kong.

14. Fortune, "Sinopec Group", https://fortune.com/company/sinopec-group/global500/.

15. Eric Ng, "China's carbon neutral goal: Sinopec plans to spend US\$4.6 billion over five years on a supply chain to promote hydrogen", *South China Morning Post*, 30 August 2021, https://www.scmp.com/business/companies/article/3146870/chinas-carbon-neutral-goal-sinopec-plans-spend-30-billion-yuan.

16. Temasek, "Temasek and blackrock launch decarbonization investment partnership", 13 April 2021, https://www.temasek.com.sg/en/news-and-views/newsroom/news/2021/temasek-blackrock-launch-decarbonization-partnership.

17. Bain & Company, "The green economy: Unlocking a more sustainable future in Southeast Asia", Press Release, 26 November 2020, https://www.bain.com/about/media-center/press-releases/2020/the_green_economy_unlocking_a_more_sustainable_future_in_southeast_asia/.

18. Sophie Yeo, "Where climate cash is flowing and why it's not enough", *Nature*, 17 September 2019, https://www.nature.com/articles/d41586-019-02712-3#:~:text=The%20UN's%20Intergovernmental%20Panel%20on,%25%20of%20the%20world's%20economy.

19. China Daily, "Top 10 best-selling electric vehicle makers", *China Daily*, https://www.chinadaily.com.cn/a/202002/17/WS5e49c4c0a310128217277e35_11.html.

20. International Energy Agency, "Global EV outlook 2020 report", June 2020, https://www.iea.org/reports/global-ev-outlook-2020.

21. Australian Trade and Investment Commission, "Annual report 2020", https://www.austrade.gov.au/international/invest/investor-updates/2020/australian-green-hydrogen-attracts-major-investment-from-japanese-giants.

22. Anmar Frangoul, "BP says Australia is an ideal place to scale up green hydrogen production", *CNBC*, 11 August 2021, https://www.cnbc.com/2021/08/11/bp-says-australia-is-ideal-place-to-scale-up-green-hydrogen-production.html.

23. CEFC, "Advancing hydrogen fund", https://www.cefc.com.au/where-we-invest/special-investment-programs/advancing-hydrogen-fund/.

24. Anmar Frangoul, "Dubai launches region's 'first industrial scale' green hydrogen plant", *CNBC*, 20 May 2021, https://www.cnbc.com/2021/05/20/dubai-launches-regions-first-industrial-scale-green-hydrogen-plant.html.

25. Paul Cochrane, "Green hydrogen: The Gulf's next big fuel — or a load of hot air?", *Middle East Eye*, 8 June 2021, https://www.middleeasteye.net/news/gulf-qatar-saudi-arabia-uae-hydrogen-green-next-fuel.

26. Jade Scipioni, "This restaurant will be the first ever to serve lab-grown chicken (for \$23)", *CNBC*, 18 December 2020, https://www.cnbc.com/2020/12/18/singapore-restaurant-first-ever-to-serve-eat-just-lab-grown-chicken.html.

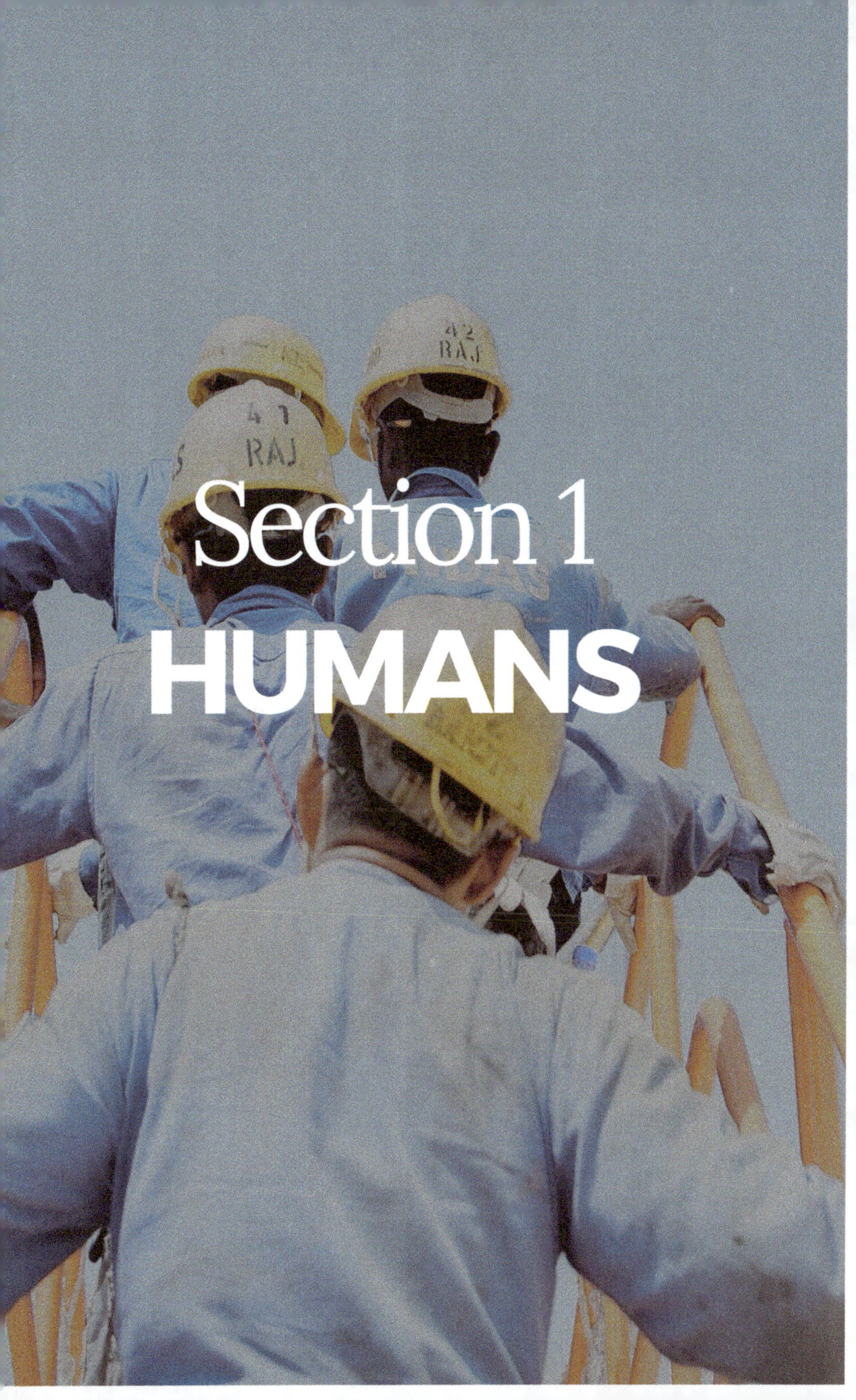

Section 1

HUMANS

INTRODUCTION

by Roman Y. Shemakov

In just 200 years, humans have reversed a 50-million-year climate cycle. If unchecked, our Earth in 2050 will be its version from three million years ago — sea levels will be 20 m higher and temperatures 3°C hotter.[1] In a way, the industrial revolution and its aftermath ushered the greatest feat of engineering known in cosmic history: a geological time machine. In accelerating the passage of planetary time, humanity has not endangered the planet; life on the "blue marble" has disappeared and reappeared for millions of years. We have mostly endangered ourselves.

The past 150 years have been an experiment in pollution. We, as a species, have been in a feedback loop of technological invention and corollary by-products. The invention of the factory invented the modern city, which welcomed the automobile, and thus the links of the modern world leapt forward through the chain of time. In the process, varieties of bio-phenomena have been eradicated, unpredictable climate disasters unleashed, and waters poisoned.

We have locked the planet into a poignant state of learned helplessness. The domestication and caricature of the environment came with an obvious pushback. But while back to the land — mid-20th century trend of escaping consumerism and corporatism via organic farming and homesteading — might have seemed like a viable option in the 1960s, we have terraformed our planet too much to simply step away. The artificial problems we have created will also need to be solved through an equally artificial plan. Technological innovation that used to be directed towards optimizing machines must now be directed towards optimizing ecosystems. A focus on by-products and externalities must move from an afterthought to the starting point. The incentive systems and consumer choices must be wholly redirected. The problem begets a need for a viable solution. With the whole world now aware of the "what" and "why", the next step is "how".

Human-led climate interference starts resembling a pharmakon: in the poison also lies the cure. The following section outlines how, in getting ourselves into this predicament, we have simultaneously proven that we can legislate, engineer, and consume our way out of it. It could be the time for humanity to completely take control of the geophysical cycle on Earth. This is perhaps more terrifying than it is reassuring, but there is no way back. The path flows through three channels: (1) reorienting multinational incentive schemes, (2) committing to massive infrastructure outlay, and (3) allowing emerging green tech firms to prioritize contribution to global environmental needs (sequestration, transparency, renewables) over short term Ventury Capital returns.

We must invent and reinvent technology that prioritizes conservation and resource recycling. We must consume the pollution we have created and embark on the first large-scale transition towards a circular geo-economy. There is value in maintaining predictable geophysical cycles. Two million years of geophysical continuity created the conditions for homosapiens to emerge, redirect rivers, form cities, mold metal, and gain access to the most fundamental physical and biological forces. The relatively stable climate of the last 7,000 years established the conditions for a flourishing human civilization and led to satellites, isotopic carbon meters, and Non-Dispersive Infrared Detectors — the tools that registered human impact on climate variations. Now, there is a possibility we can destroy the environment that produced the ability to interpret an environment.

The first chapter "Policy & Finance" by Alexandra Tracy looks at the tools we must employ to bring a Carbonless future into reality. First, the chapter outlines the history of green finance. It focuses on the effort of the Paris Climate Agreement and regional banks to create national decarbonization strategies. One of the most important developments over the past 20 years has been the role of large investor alliances pooling money around Environmental, Social, and Governance (ESG) goals together. As the section's narrative moves towards the present, the appetite for green investment has increased significantly. The chapter highlights important trends from the People's Bank of China, Japan's Mitsubishi UFJ, HSBC, and Macquarie's emerging emphasis on green infrastructure. As we look ahead, it will become even more important to measure and embrace our responsibility to future generations and reposition global portfolios in line with that vision. Carbon Pricing, Shareholder Advocacy, Green Debt

Markets, and specialized institutions will become vital to reach humanity's most important goal this century.

In the second chapter "Industry & Waste", Sandro Desideri analyzes the industrial production and destruction cycle ("cradle to grave"). We look at the outsized effect of the heavy, manufacturing, and transformation sectors on the global Greenhouse Gas (GHG) output. Currently, overall industrial activities contribute more than one-third of anthropogenic GHG emissions, while the heavy industry (steel mills, cement plants, the construction sector) is originating 10% of the global GDP and more than double (22%) of the GHG. The trends in Combined Heat & Power, Green Hydrogen, Renewable Maximization, Microgrids, and Waste Heat Recovery have the potential to alleviate a large portion of the destructive by-products. Unfortunately, they are far from being fully embraced across the industrial supply chain. Thoughtful international frameworks must set ambitious targets but, in turn, work to alleviate transition costs and prevent "carbon leakage".

The third chapter "Lifestyle & Consumption" by Christine Loh investigates the impact of consumer decisions on the global fight against (or for) climate change. Over the past century, tastes have changed radically, and the industries alongside them, though not always in that order. This chapter analyzes how green consumer choices have given birth to new industries like plant-based protein, electric vehicles, and urban recycling. As more actionable data, decentralized monitoring, and advanced carbon accounting become widespread, consumers will become increasingly more exposed to the embodied carbon in their products, and green consumer decisions will more accurately reverberate onto the producers. Consumer education, transparent industries, and flexible supply chains will be the key for buyers to lead Asia towards Decarbonization.

Endnote

1.   Susan Scutti, "In 200 years, humans reversed a climate trend lasting 50 million years, study says", CNN, https://www.cnn.com/2018/12/10/world/climate-change-pliocene-study/index.html.

Chapter 1
POLICY & FINANCE

by Alexandra Tracy

ANALYSIS

by Alexandra Tracy

Like elsewhere around the world, decarbonization in Asia-Pacific is wholly dependent on the availability of finance. The Asian Development Bank (ADB) forecasts that emerging Asia will need to invest close to USD 1.7 trillion per year to 2030 on developing low carbon infrastructure and meeting climate goals.[1] Achieving continued economic growth in the region while meeting decarbonization targets will require ambition and focus by policymakers, regulators, and the entire financial industry. Significant change is expected in the following areas:

- reducing capital flows into high emitting sectors,
- allocating more funding to low carbon and transition businesses, and
- mobilizing investment at much greater scale in emerging markets.

PAST

The United Nations Framework Convention on Climate Change has been an important driver of awareness for governments across the world. The global commitment in 2010 to invest USD 100 billion per year in emerging markets,[2] followed by the Paris Agreement in 2015 (requiring countries to submit investible climate action plans),[3] have put funding strategies at the top of the agenda. These events are considered milestones in the emergence and growth of sustainable finance historically.

Green Finance is Born

The Group of Twenty heads of state in 2016 emphasized the need to scale up climate and environmentally sensitive investments and embed sustainability within the financial services industry.[4] A Green Finance Study Group was set up to encourage private investors to increase green investments.

At the same time, governments in Asia started to establish national frameworks and strategies to support green finance. The People's Bank of China (PBOC) convened a Green Finance Task Force in 2015, which proposed policy support for green banks, funds, and market infrastructure enabling green bonds, loans, and services.[5] In Indonesia, the Financial Services Authority launched a "Roadmap for Sustainable Finance in Indonesia",[6] which called for regulatory incentives, design of specific products, and greater expertise for financial institutions.

Supportive Capital Market Regulation

Securities regulators and stock exchanges across the region also implemented new sustainability requirements and reporting regimes for listed companies. Bursa Malaysia began the trend in 2006 by asking companies to report on their corporate social responsibility efforts.[7]

Both the Singapore Exchange and the Hong Kong Stock Exchange (HKEx) have been regularly upgrading their environmental and social reporting frameworks, with an increasing emphasis on climate, such as incorporating the reporting methodology developed by the Task Force on Climate Related Financial Disclosures.[8]

Integrating Sustainability into Investment

Another key element of capital market development is the behavior of the largest investors, who have huge influence over their fund managers and portfolio companies. National Pension Service in South Korea was one of the first in Asia to allocate capital to a sustainable portfolio, and has been followed by many of its peers.[9]

Asset owners have also joined investor alliances seeking to encourage greener investing. The Government Pension Fund of Thailand was one of the founders of the United Nations (UN) Principles for Responsible Investment, which promotes the integration of environmental, social, and governance (ESG) considerations into investment decision-making.[10] Many are also members of the Asia Investor Group on Climate Change, a forum for examining risks and opportunities associated with low carbon investment in Asia.[11]

PRESENT

There have been strong momentum and enormous growth in demand for climate investment products over the last two years. The net zero commitments announced by the region's biggest economies — first in 2020 by China, Hong Kong, Japan, and South Korea, and more recently by India, Saudi Arabia, Vietnam, and Thailand have been important policy signaling exercises, directed at least partly at the financial markets.

Capital is There

Public balance sheets were already stretched very thin, even before the response to COVID-19. For example, the PBOC has emphasized that China's government can finance only 15% of the annual investment needed to meet its environmental targets, while the rest must be met by private sources.

But the capital is there — global assets under management grew by 40% in the past five years to USD 110 trillion.[12] In turn, investors are signaling that they are ready to make significant progress on these issues, with the support of strong and consistent policy frameworks. Together with their global peers, several of the region's largest financial institutions, such as HSBC[13] and Japan's Mitsubishi UFJ,[14] have adopted their own net zero targets.

Rapid Growth of Capital Flowing into Sustainable Investments

ESG funds attracted unprecedented inflows in 2021, prompting a rush of new fund launches. Globally, sustainable investment assets have grown 15% over the past two years to USD 35 trillion,[15] which represents 36% of all professionally managed assets. Japan's market is the most advanced in Asia as of 2021, with allocation to sustainable assets at 24% of the total market.[16]

As shown in Figure 1.1, funding into climate technology is accelerating. Over USD 23.2 billion in listed equity capital was raised for sustainable businesses in the first half of 2021, with Asia accounting for 38% of

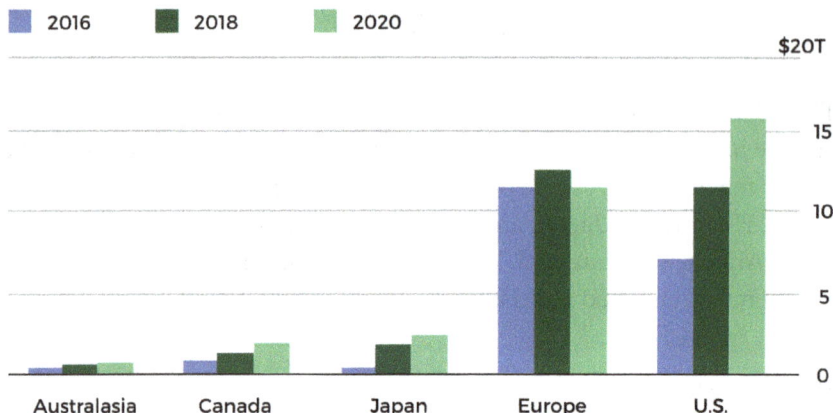

Figure 1.1.   Global venture capital flow into climate tech.

Source: Global Sustainable Investment Alliance, "Global sustainable investment review 2020".

the total.[17] Issuance of green bonds was also up sharply, reaching nearly USD 260 billion, of which around 20% was in Asia.[18] Even mergers and acquisitions of sustainable firms were at a record high during the same period, with dealmaking worth USD 84 billion. Over a quarter of this activity involved China.[19]

Appeal of Infrastructure

There is now widespread recognition of infrastructure as an important sustainable asset class, which is attracting increasing capital flows. Preqin forecasts that unlisted infrastructure assets under management will grow to USD 795 billion in 2025.[20] A recent survey by Macquarie Infrastructure and Real Assets found that over 91% of a pool of more than 150 of the largest institutional investors plan to increase their focus on ESG principles in the next five years.[21]

Macquarie, the world's largest infrastructure manager, signaled its own commitment to decarbonizing its operations in 2017 with the acquisition of the United Kingdom's Green Investment Bank, a leading investor in green infrastructure in Europe.[22] Its Asian portfolio now includes around 9 GW of renewable energy capacity under development.[23]

FUTURE

As climate risk becomes more critical, and the opportunities associated with decarbonization more visible, executing on countries' net zero promises calls for rapid acceleration of capital flows into decarbonization and a radical repositioning of investment portfolios.

Restricting Capital Flows to High Emitters

Regulators are intervening more directly to increase the cost, and reduce the total amount, of capital available to finance the activities of high emitting businesses, as financial institutions step up efforts to decarbonize.

Improving Disclosure and Tackling Greenwashing

Investing in support of a low carbon transition requires credible and consistent data about the assets being financed. As the sustainable investment industry has expanded, there has been a proliferation of data providers, rating agencies, analysts, and standards setters aiming to increase transparency and hold companies and financial institutions accountable.

In Asia-Pacific, the number of companies reporting Scope 1 and 2 emissions has grown steadily since 2015. The fastest growth is in Japan, Singapore, Malaysia, and India, largely driven by greater expectations from local stock exchanges and regulators.[24] In Japan, the Government Pension Investment Fund (the largest asset owner in the world) has also played a major role in influencing local asset managers to improve disclosure.[25]

As countries in the region move away from voluntary disclosure to mandatory reporting regimes, they are developing their own national standards. For example, Malaysia is implementing a principles-based taxonomy for banks, which will require them to report the carbon exposure of their loan books.[26]

There has also been greater scrutiny of fund managers and their portfolios. In Hong Kong, the Securities and Futures Commission recently issued guidance on enhanced disclosure by funds claiming to be green or sustainable, with a particular focus on climate data.[27]

As ESG data becomes more complex, a number of technology start-ups are developing platforms that can gather information from multiple sources and improve reporting and analysis. Diginex, a global blockchain financial services and technology company, is looking at ways to use its expertise to improve ESG reporting.[28] A local example is RIMM Sustainability in Singapore, which provides tools for sustainability disclosure and optimization to both companies and fund managers.[29] *(Full disclosure: the author is a non-executive director of RIMM Sustainability Pte Ltd.)*

Using Carbon Pricing to Allocate Capital

At meaningful levels, carbon prices will play an important role in making the business case for decarbonization. As carbon pricing schemes and carbon markets are introduced around the world, they impact the cash flows of companies, especially large emitters, and are increasingly having an effect on corporate valuations.

There are 64 carbon pricing mechanisms in operation globally, covering 21.5% of global GHG emissions.[30] China's national carbon emissions trading scheme was launched on the Shanghai Environment and Energy Exchange in July 2021. It will become the world's largest carbon market, initially covering close to 2,200 power plants, which already account for around 12% of global carbon dioxide emissions.[31]

Even in jurisdictions where carbon pricing schemes are not yet in place, companies are beginning to use internal carbon pricing to assess the possible financial impact of their emissions (visualized in Figure 1.2). In reporting to CDP, an international non-profit that helps companies to report their environmental impacts, more than 2,000 companies this year disclosed such activity.[32] Temasek Holdings Limited (Temasek), a Singapore government investment company, has also set an internal carbon price of USD 42 per tonne as part of its strategy to incorporate climate factors into investment decisions.[33]

MioTech, a Hong Kong startup, which uses artificial intelligence to help companies with tracking carbon emissions and monitoring ESG performance, expects opportunities in China to accelerate after the Shanghai

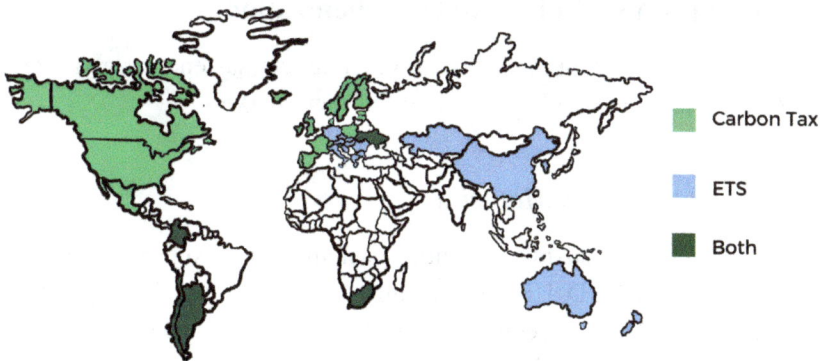

Figure 1.2. Carbon tax and ETS around the globe.

carbon exchange has gone live. The company is planning to double its headcount in the region to 300 people.[34]

Shareholder Advocacy

Asset owners and their fund managers now regularly engage with companies on carbon risk and decarbonization strategies. Engagement can escalate from requesting greater clarity on portfolio performance to voting on corporate resolutions and, ultimately, divestment. Larger groups of shareholders are also putting pressure directly on portfolio companies. For example, Climate Action 100+, supported by more than 615 investors controlling over USD 55 trillion,[35] targets the world's largest emitters, including 34 Asian companies.[36]

Investors are also targeting other financial institutions in efforts to reduce the flow of capital to high emitters, with considerable success. A number of Asia's largest banks, including HSBC,[37] Sumitomo Mitsui,[38] and DBS Bank,[39] are phasing out financing for the coal industry. Similarly, in the insurance industry, regional players such as MS&AD in Japan[40] and Hyundai Marine & Fire Insurance and Hana Insurance[41] in Korea have announced plans to reduce the availability of cover for coal-fired power plants.

Increasing Capital Flows to Decarbonization

Growing demand from investors for green assets has encouraged product innovation and the rapid development of new markets.

Green Debt Markets Maturing

Green, social, and sustainable bond issuances in Asia in the first half of 2021 totaled USD 100 billion, already surpassing the total for the year earlier.[42] As the green bond market matures beyond simple issuances of securities by highly rated corporate and public sector entities, issuers are looking to raise capital through a variety of tailored fixed income instruments to support decarbonization. These include project bonds for infrastructure, environmental impact bonds, and asset-backed securities.

Asia's first certified climate bond was a USD 225 million project bond, backed by the ADB, financing the Tiwi-MakBan geothermal energy facilities in the Philippines.[43] A partnership between Singapore government-backed Clifford Capital and the Asian Infrastructure Investment Bank is currently working on packaging energy loans,[44] while the Hong Kong Mortgage Corporation has announced a plan to securitize infrastructure assets.[45]

The market has also grown rapidly for green loans, whose interest rate is tied to the borrower's sustainability performance. In Asia, excluding Japan, a total of USD 22 billion was committed in the first half of 2021.[46] Borrowers are as diverse as Leo Paper in Hong Kong,[47] Thai Union, one of the world's largest seafood companies, and China Mengniu Dairy.[48]

Transition Funding via Equity and Direct Investing

According to Bloomberg, investment in the energy transition globally grew to USD 501 billion in 2020.[49] Investors continue to commit capital at scale: the Brookfield Global Transition Fund (backed by Temasek and several Canadian pension funds), for example, reached its first close in July 2021 at USD 7 billion.[50]

The world's largest asset manager, Blackrock, also sent a strong signal to investment markets by announcing in April 2021 that it will work with Temasek to launch a series of late-stage venture capital and early growth private equity funds focused on advancing decarbonization solutions. The partners will themselves commit USD 600 million to seeding the funds.[51]

Climate tech investors are looking beyond renewable energy. Transportation was the dominant sector in 2020, accounting for more than half of all investment deals by value.[52] Climate solutions in heavy industry sectors, food, agriculture, and land use are attracting increasing amounts of capital.[53]

The investor community is also becoming more diversified. In 2019, around a third of all deals in transportation and energy included a corporate investor.[54] In addition to capital, these participants bring industry knowledge and commercial experience that can help startups deploy and scale up their innovations.

The number of new-age cleantech or climate tech (as shown in Figure 1.3) — the terminology varies — venture capital funds based in Asia is growing. Recent entrants include Singapore-based DreamLabs Innovation, a USD 50 million fund looking to back cleantech, energy, and digital technologies.[55] The Razer Green Fund in Hong Kong, also USD 50 million, aims to invest in environmental and sustainability startups in the seed and Series A stages.[56]

Rise of Natural Capital Investment

Opportunities for investment in natural climate solutions (using natural ecosystems to reduce or remove emissions) are receiving increasing attention both from businesses seeking to reduce risks within their supply chains and from financial institutions.

These solutions can provide as much as a third of the cost-effective climate mitigation needed globally to achieve net zero emissions by 2050,[57] but the pace of investment needs to accelerate rapidly. According to the UN, a USD 4.1 trillion funding gap must be closed by 2050.[58]

Investing in natural capital takes place in multiple ways, from upfront investment in project development to the purchase of carbon credits

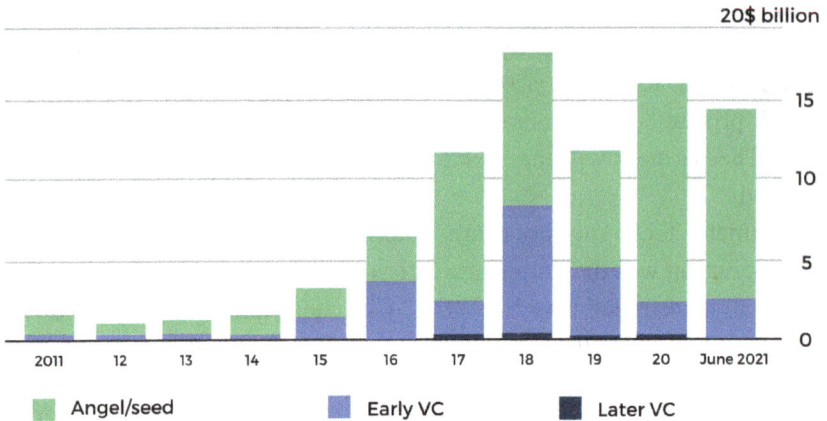

Figure 1.3. Global venture capital flow in climate technology.
Source: Pitchbook.

through the voluntary carbon market. Global fund managers, such as Althelia Ecosphere (now part of Mirova),[59] are investing in areas such as forestry, conservation, and sustainable farming.

In Asia, the Meloy Fund for Sustainable Small Scale Fisheries focused on Indonesia and the Philippines and recently attracted investment from foundations and family offices.[60] There is also significant potential for investing in ecosystem solutions in the South Pacific islands, especially to support adaptation measures, but mobilizing capital at scale is challenging, and this region is still very dependent on overseas aid funding.

As the carbon offset market expands and potential new products emerge, such as blue carbon (representing carbon sequestered in marine ecosystems), expertise is needed to create methodologies for measuring and verifying carbon savings created by different technologies. This is producing opportunities for technical consultants, such as Carbon Direct, which is advising on projects around Asia.[61]

Mobilizing Capital at Scale in Emerging Markets

Some domestic capital markets are coming up rapidly, but most emerging market economies still require significant amounts of foreign capital. In these markets, the banking sector tends to be dominant, and there is often a large informal sector, including agricultural cooperatives, credit unions, and microfinance.

Catalyzing Investment with Blended Finance

Investing in emerging markets is often challenging, especially for cross-border investors who are unfamiliar with local conditions. Development banks and multilateral institutions, such as the World Bank, have a long track record of catalyzing private investment in viable projects by employing the techniques of blended finance. A group of funders is brought together into different investment tranches, each taking the level of risk and return acceptable to them.

In addition to providing concessional funding to emerging markets from its Climate Investment Funds,[62] the World Bank works directly with the private sector via its sister organization, the International Finance Corporation (IFC). In Asia, the IFC supports climate investments directly, but also operates through local financial institutions, enabling them to become active in these sectors. For example, its IFC China Utility Based Energy Efficiency Programme has catalyzed USD 790 million worth of loans to over 220 projects, reducing the equivalent of 19 million tons of carbon dioxide annually.[63]

The UN Green Climate Fund, based in Seoul, has mobilized funding for more than 150 projects in emerging markets by providing concessional funding and de-risking mechanisms to encourage co-investment. It has supported investments in the region, from energy saving in manufacturing in Bangladesh to reducing emissions through sustainable landscape management in Laos.[64] It is also active in the South Pacific, directing about USD 1 billion of funding into 38 projects, including climate-resilient urban water infrastructure in Fiji,[65] renewable energy in Tonga,[66] and coastal protection in the Marshall Islands.[67]

Blended finance transactions often require a combination of creative financial engineering and broad partnership. The Tropical Landscapes Finance Facility in Indonesia provides long-term financing for renewable energy and forestry-based smallholder livelihood projects. It was put together by a grouping of fund manager ADM Capital in Hong Kong, the

UN Environment Programme, the World Agroforestry Centre, and BNP Paribas, which will repackage the resulting loans into a medium-term note programme.[68]

Specialist Institutions for Green Finance

Efforts to mobilize funding for emerging markets have received a boost from the emergence of "green investment banks", dedicated financial institutions with a mandate to raise funding for green infrastructure. They employ customized financing tools to bring in the private sector and thereby create liquid markets for low carbon technologies.[69]

For example, the Malaysian Green Technology and Climate Change Centre (formerly GreenTech Malaysia) was set up in 2010 to develop renewable energy and green technology markets in the country. It has provided technical assistance and loan guarantees to nearly 30 local banks and financial institutions. Its scope was recently expanded to cover climate change mitigation and adaptation.[70]

Policymakers around the region, including in the Philippines, Thailand, Vietnam, and Indonesia, are developing plans for their own national green investment banks. The Mongolian government recently announced the Mongolian Green Finance Corporation, which will mobilize local and international capital into climate change sectors, initially through energy efficiency and green housing projects in the country.[71]

CONCLUSION

Focus on climate by policymakers and investors across Asia has continued to accelerate since the first net zero commitments from regional governments in 2020. At the recent United Nations Climate Change Conference (COP26), Asian leaders pledged their support for international agreements on scaling back coal use, reducing emissions caused by methane, and protecting forests.

COP26 strongly emphasized the need to mobilize private finance to meet climate goals. At the conference, the Glasgow Financial Alliance for Net Zero of over 450 financial institutions, responsible for assets worth USD 130 trillion, committed its members' support for the transition to net zero.[72] The unprecedented scale of this undertaking reflects a clear understanding that this global effort requires action from the mainstream private finance community and buy-in from the entire financial system in both developed and emerging markets.

In Asia, this will require further efforts to put in place supportive regulation and to improve industry-wide metrics and standards for green financing, as well as more education of the private sector about the opportunities to be gained from sustainable investments. While each country faces its own individual challenges in moving its financial industry away from business as usual, there is enormous potential for partnerships across the region, sharing of best practices, and greater alignment in order to achieve long-term investment goals.

STARTUP CASE STUDY
Allinfra

by Alexandra Tracy

Poor Quality Climate Data Undermining Green Investment

One of the greatest challenges associated with sustainable investing is the collection of accurate data. As more companies around Asia announce decarbonization strategies and net zero commitments, investors and regulators are looking for effective ways to measure their performance and hold them to account. The rapid development of carbon markets and systems in the region will further accelerate the demand for more sophisticated reporting on emissions reductions.

Reliable data collection is essential not only for corporate reporting but to give credibility to ESG ratings and green finance instruments, such as green bonds and loans. It is also essential to support the creation and trading of new environmental financial products, such as emissions reduction certificates, which are critical to meeting climate goals in the region, by allowing organizations in high emitting sectors to offset their carbon footprints in the short term by funding activities that avoid, reduce, or sequester carbon.

However, the traditional approaches to monitoring emissions reductions and the environmental impact of projects are beset by a number of weaknesses. Data gathering still tends to rely on manual processes and is labor-intensive and time-consuming. The periods between measurement dates can be significant, undermining the ability to predict output volume and timing. Crucially, the output may not be transparently tied to the underlying environmental data, which leads to concerns about double-counting and uncertainty about the actual environmental benefits.

Solving these problems will be crucial to creating greater investor confidence and scaling up the appetite for environmental financial products in Asia.

Improving Data Management with Blockchain Technology

Established in 2019, fintech startup Allinfra Ltd aims to transform the renewable infrastructure and environmental financial product markets with technology that ensures transparency and availability of real-time energy and emissions data. With backing from ConsenSys, one of the world's leading blockchain software companies, Allinfra's mission is to revolutionize the way that clean assets are monitored, traded, and financed. Allinfra is led by a team of industry experts with experience across capital markets, renewables, infrastructure, and technology.

Allinfra Climate

The company's flagship product, Allinfra Climate, is a blockchain-based environmental platform that collects climate-relevant information directly from the source, recording data from smart devices, building management systems across an organization's portfolio of assets, and creating a permanent repository that can be easily audited.

The use of blockchain simplifies what has typically been an administratively heavy, manual process and creates a way to track and store information that is highly trustworthy and gives greater certainty to users about the source and specifics of the outputs. Information can be made available on the underlying infrastructure asset, the specific meter, and reporting time period, reducing the likelihood of double-counting or misrepresentation. Real-time digital data gives a greater understanding of outputs over time and allows for better planning of future production.

Distributed ledger technology means that all recorded transactions and information are shared across a network, with each point in the network validating the authenticity of changes. No third parties are involved in reporting data on the underlying asset to the holders of any digital instruments linked to it, which reduces costs and potential for errors. Similarly, instruments can be traded or retired directly by the owners, with no need for financial intermediaries.

Asset	Smart meter	Allinfra technology	Allinfra Climate	Financial services		

Data analysis — Internal, external reporting and climate risk management

Financial services — Source data for ESG ratings, green financing (green bonds, loans, sustainability linked instruments) and more

EFPs — Underlying data source for a variety of financial products like renewable energy certificates and emission reductions

① Energy usage and other environmental data is recorded by existing devices and sensors connected to the asset

② Allinfra's technology securely and auditably takes data direct from the device and prints it to the blockchain

③ Asset owners can utilize this data for multiple purposes without requiring verification or validation from an intermediary

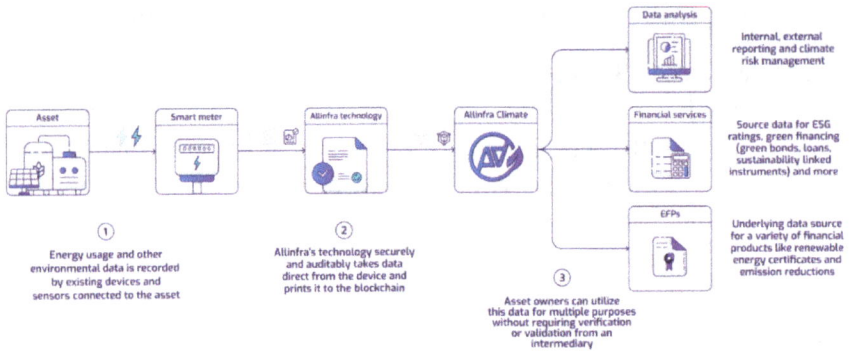

Figure 1.4. An end-to-end solution allows users to store, use, or monetize data.

The Allinfra Climate platform (explained in Figure 1.4) can be used to enable internal and external corporate reporting, improve energy management, and mitigate environmental risks. When combined with smart contracts, stored data provides the backing for green financing instruments and can underpin the creation of a variety of digital environmental financial products, from renewable energy certificates to emission reductions and associated derivatives. The technology can be implemented across many asset types and industries, including commercial real estate, infrastructure, renewable energy, and vehicles.

Two of Asia's leading renewable energy companies are currently using the Allinfra Climate platform to capture data from selected assets within their portfolios. UPC Renewables, one of the largest independent renewable energy companies in the region, has wind and solar plants totaling over 6 GW in operation, under construction, and in development. At the micro-level, Canopy Power serves customers throughout Southeast Asia and the Pacific who either have no mains electrical grid access or access only to a weak electrical grid. Allinfra's technology is being employed by Canopy Power for a number of projects in Malaysia and Indonesia.

Creation of Digital Renewable Energy Certificates

A major thrust behind the development of the Allinfra Climate platform was to mobilize financing for clean infrastructure and renewable energy assets in Asia. For some years, investors have been able to fund these activities by purchasing renewable energy certificates, which represent the environmental attributes associated with the projects, such as emissions reductions. Through the platform, Allinfra is able to create a digital Renewable Energy Certificate (dREC), which has all the rights and attributes of traditional RECs, but offers greater efficiency and transparency by tracking and storing all relevant assets and certificates information. Every certificate is permanently linked to underlying energy generation data from a specific recording device.

A dREC can be held, traded, or retired (meaning that the certificate is canceled and can no longer be sold) within Allinfra Climate's renewable energy certificate trading platform. Transfer from party to party or the retirement of a dREC is recorded on the blockchain, ensuring that any claim of environmental impact can only be made once, and permanently recording the retirement of such a claim.

Future Developments

In 2021, Allinfra closed a strategic funding round from a broad group of aligned investors, including UPC Capital Ventures, BC Group, FJ Labs, ACRE Investment, ConsenSys, and the significant shareholders of several Asia-Pacific renewable energy platforms. This funding round will enable Allinfra to accelerate product development across infrastructure, property, and industrial assets and evolve its offering in forestry, agriculture, and land use space.

Allinfra is also involved in a number of research and project collaborations aimed at using blockchain technology to improve climate reporting and facilitate green finance. For example, it is working with KPMG on the development of its blockchain-based Climate Accounting Infrastructure, which is intended to help organizations more accurately measure, mitigate, report, and offset their greenhouse gas emissions. It is also a partner, together with the Bank for International Settlements Innovation Hub and the Hong Kong Monetary Authority, in Project Genesis, which will use blockchain to drive transparency of green investments and increase the ability of small investors to participate through the tokenization of assets.

CORPORATE CASE STUDY
HSBC Group

by Alexandra Tracy

Asia's largest capital providers exert enormous influence over the economy and its potential to achieve rapid decarbonization. In countries where the capital markets are underdeveloped, in particular, banking institutions are the dominant players, which have the power to channel a large proportion of the available funding to businesses.

HSBC Group is one of the region's largest banks, enjoying significant market share in wholesale, retail, and investment banking and controlling assets of nearly USD 3 trillion.[73] While the group has substantial operations globally, its flagship businesses are in Asia, which during the first half of 2021 generated 64% of HSBC's global pre-tax profits of more than USD 10.8 billion.[74] Importantly, HSBC has a significant presence in most emerging markets in the region, with a longstanding and diverse client base.

HSBC's Net Zero Strategy

Building on existing initiatives and targets that were implemented (and increased subsequently) after the Paris Agreement in October 2020, HSBC launched an overarching strategy to support the global transition to a net zero carbon economy by increasing the amount of sustainable capital available to its customers to achieve their own carbon reduction goals. The strategic shift has resulted, amongst others, in:

- Providing between USD 750 billion and USD 1 trillion in sustainable financing and investment by 2030.
- Working with partners to innovate and increase investment in natural resources, clean technology, and sustainable infrastructure.
- Reducing its own emissions from operations, supply chain, and within the portfolio to net zero by 2050.

This strategy was approved as part of a special resolution on climate change at HSBC's Annual General Meeting in May 2021.[75] Implementation of the strategy focuses on a number of key deliverables:

1. *Withdrawal of Financing from High Emitting Sectors*

- HSBC will implement a policy to phase out the financing of coal-fired power and thermal coal mining by 2030 in the European Union and developed countries, and by 2040 globally. This commitment requires the bank to transform its portfolio significantly, as it is currently a large funder of fossil fuels in Europe and works extensively with the coal sector in Asia.

- In the palm oil and agricultural commodities sectors, HSBC maintains its policy of requiring sustainable certification from companies and will not provide financing to any businesses that do not comply with "No Deforestation, No Peat, and No Exploitation" standards.

2. *Increased Funding for Decarbonization and Transition*

- HSBC is one of the largest providers in Asia of sustainable finance solutions, including green bonds, social and sustainability-linked bonds, and loans. From January 2020 until the end of the first half of 2021, the bank facilitated USD 87 billion of sustainable finance and investments.[76]

- Nearly 89% of all assets under management in HSBC's fund management business were invested according to sustainability principles as of December 2020.[77] A series of lower carbon equity and bond funds are distributed to consumer and private banking clients. The bank also recently introduced its first private equity impact fund in Asia.[78]

- HSBC plans to expand its technology venture debt capabilities to provide USD 100 million of financing to companies developing clean technologies in the energy, transportation, insurance, agriculture, and supply chain sectors. In 2021, the bank announced the appointment of a new "Climatech" team, with the first fund to be launched by the end of the year.[79]

- HSBC has also introduced green products specifically targeted at making accessing financing easier for SMEs in the region. For example, in 2020, it launched an SME green loan scheme in Singapore that reduces the time, complexity, and cost typically associated with applying for green finance.[80]

3. *Product Innovation and Partnership*

- HSBC is leveraging its expertise in emerging markets as part of initiatives to accelerate capital flows into climate-smart infrastructure in these regions. For example, it partnered with the International Finance Corporation to establish the Real Economy Green Investment Opportunity Fund (REGIO)[81] and is a leader in the "Finance to Accelerate the Sustainable Transition Infrastructure" initiative with the World Bank and others.[82] In September 2021, a separate venture was announced with Temasek to catalyze the financing of marginally bankable sustainable infrastructure projects in Southeast Asia.[83]

- In a response to the growing demand for nature-based solutions to climate change, HSBC created a joint venture in 2020 with specialist climate change advisory and investment firm, Pollination, which aims to be the world's largest dedicated natural capital asset management company.[84]

- Climate Solutions Partnership is an initiative with the World Resources Institute and the World Wide Fund for Nature, backed by USD 100 million of funding from HSBC, which aims to catalyze climate innovation ventures and nature-based solutions and help to transition the energy sector towards renewables in Asia. Using the WWF's Impact collaboration platform, the partnership will support business innovations to scale, in collaboration with universities, research institutes, and incubators.[85]

4. *Heightened Expertise and Best Practice*

- HSBC's Centre of Sustainable Finance (previously the Climate Change Centre of Excellence) was established in 2010 to provide research and thought leadership on the financial system response to climate change risk. Building on internal subject matter expertise and an external network of experts, it looks at low carbon transition solutions for the energy system and infrastructure and aims to disseminate information and best practices.[86]

- In 2020, HSBC set up a dedicated ESG Solutions Unit to provide advice, expertise, and sustainable financing ideas to clients around the world and to launch new products and asset classes. The team is expected to expand over time to meet the growing requirements of businesses transitioning towards net zero carbon.[87]

- The bank is committed to raising the level of expertise with sustainable finance among its business units more widely. A program was recently launched with the Chartered Banker Institute in the UK to provide e-learning on green and sustainable finance to more than 500 staff members over the next year.[88]

INTERVIEW
Dr. Steve Howard
Chief Sustainability Officer, Temasek

by Alexandra Tracy

Dr. Steve Howard is Chief Sustainability Officer of Temasek, the Singapore investment company with over SGD 380 billion in assets.[89] Previously, he was Chief Sustainability Officer at the IKEA Group. He is also co-chairman of the We Mean Business Coalition and an advisory board member of Sustainable Energy for All.

Investing in decarbonization is a huge topic. Where do you start?

You've clearly got areas such as energy, transport, food, and agriculture where the best growth areas are green growth. If you're looking at extending the power grid, decentralized renewable energy may be the lowest cost option as well as the cleanest and greenest option. Similarly, vehicle electrification is now coming to cost parity or better. So are developments in smart industries and vertical farming in a city context. All of this is creating win-win investment opportunities.

Are investors already starting to see the big picture?

We tend to look backward and not forward. You've got multiple, exponential curves where you can see technology scaling and costs coming down. As we go into new areas, people will price future risk differently from historic risk, but we're much better at saying next year will be incrementally different from the last year than we are at looking at transformational, sweeping developments. You have to look forward, focus on the breathtaking pace of change, and lean into that.

How do investors keep on top of these kinds of sweeping changes?

Investors need to be super savvy and educated about this because these changes over one or two or three decades will totally transform food and agriculture, built environments, mobility, and energy. These are non-trivial developments. It looks very similar to digitalization in terms of the way that it will also transform industries and business models. We're now into the distribution phase and seeing tremendous business model innovation based on clean energy and green systems. From an investor's point of view, it's very exciting.

You seem very confident that costs are reducing across sectors?

For the purposes of planning, assume that everything will get cheaper. There is now so much capital globally starting to go into the rapidly evolving sectors like green cement, green steel, and green hydrogen. Let's take the most aggressive and optimistic projection of how cheap they are going to get, and that will probably be the right one.

What about the price of carbon?

Everything will get cheaper — except carbon. That's the only thing that's likely to get more expensive. Policymakers are going to be using all tools in the toolbox to squeeze carbon out of the economy, so investors need to recognize and respond to this. At Temasek, we have implemented an internal carbon price of $42 a tonne. We are going to be increasing that year on year.

Isn't aggressive carbon pricing going to hurt many businesses and their investors?

For us, as a long-term investor, taking a 20-year view on our returns, aggressive policy action on climate change might have a slightly depressive effect in the short term, but in the long term, it's much better. A world in which we invest aggressively in tackling climate change gives us much more positive long-term returns than one where we don't. If you're an investment institution that has some scale, you have to play a part in this.

What does your own portfolio transition look like?

We've said we're going to halve our portfolio emissions by 2030 against 2010 levels — that's more than half of where we are today — and we have a long-term goal for net zero by 2050. The net zero is important, but the 2030 target is more important because it drives immediate action.

What's the roadmap for getting there?

First of all, you need to understand where the emissions are in your portfolio. We've got a pretty good picture of that. We certainly know where the chunky emissions are. And we are sending a strong signal to our portfolio companies about what our views are on this. As a long-term investor, we want to work with them on transition, by sharing knowledge, looking at the technology solutions, and working across the ecosystem to look at how we get to these sorts of goals together.

Is there a role for carbon offsets in your strategy?

Strategy one is to reduce emissions across the portfolio. That's absolutely the headline strategy. But in 2030, there will definitely be residual emissions, so we will use offsets in all likelihood to neutralize them. We want to make sure most of the progress will be from real reductions, which also have a business benefit (greater efficiency or lower energy costs), as opposed to offsets, which are just an expense.

How do you see the offsetting market developing?

High liquidity, high trust carbon markets are really important, especially in getting capital flowing into nature-based solutions: regenerative agriculture, regenerative forestry, avoided deforestation, etc. These can be 20% or more of the solution to tackling climate change. Temasek has helped to set up the Climate Impact Exchange in Singapore to create a transparent market place for nature-based offsets. I hope that initiatives like this, as well as the Task Force on Voluntary Carbon Markets, will be able to create a hardcore of trust for investors.

Is the technology out there to support large-scale decarbonization?

We've been investing in relatively early-stage technology for some time. But we are also looking at opportunities really to scale technologies as they mature and become proven. We have created Decarbonization Partners, our new joint venture with BlackRock, explicitly to address that segment. The pipeline of potential investments for us is super exciting. We will be setting up a series of funds and working with other investors to explore how we use our networks to drive scale, find the right channels to market, help navigate challenging industrial systems, and grow successful enterprises rapidly.

Do you see more appetite for these sectors from your institutional peers?

We do see other like-minded investors in the space now. We're also partnered with Ontario Teachers, for example, a big Canadian pension fund, with Brookfield to set up the Brookfield Global Transition Fund, which looks at how to use capital to deploy renewables and transform brown industries into green. In fact, I think there's been a fast awakening. If you had asked me two years ago, I would have said capital markets are largely asleep. Now capital markets have woken up, had a coffee, and got to work.

What do you think has energized the wider market?

We saw the science advance and the IPCC come out with their 1.5°C report. It was a bit of a wake-up call, which was really saying we've nearly blown the budget to stay under 1.5°C, and 1.5°C is the upper limit of where we should go. The social movements that have happened have been an accelerant because they push policymakers; the activists, the school climate strikers, and the Greta Thunberg effect have been material in this. Most major governments around the world now have a net zero target, mostly 2050. Then we've seen 2,000 or so corporate commitments on science-based targets, net zero commitments, 100% renewable energy, etc. And there's the extreme weather that we see around the world. Wake up and smell the forest fire!

So now is the time for action?

All of these things have come together, and at the same time we've seen green systems maturing so rapidly. We're very fortunate in that this is something we can solve with technology that's largely to hand. With some technologies, we need to drive innovation and investment, but actually, a lot of the solutions are ready to scale. This clean revolution is embedded and durable, and it's not going away.

But there are still some challenges along the way?

We still need to look at how do we do the hard to do stuff — where you've got assets that one day will become stranded because they are carbon-intensive, but they're not stranded maybe for a decade or so — and we need to be really planning the transition for those heavy industries, hard-to-abate sectors. And we need to think about how we finance the hard-to-finance projects — where the risks and rewards are challenging, where you have to do extra work, more due diligence, and potentially have some blended finance in the mix. We've recently set up a debt platform with HSBC to do sustainable finance. We're a partner with LeapFrog, which is one of the pioneers of impact investing and related reporting.

Looking back at your time at IKEA, do you see any differences of approach when compared to an institution like Temasek?

Large investors tend to have more complexity. You have to look at how you can adopt approaches that can work across many, many investments. But in some ways, you've got more nimbleness as an investor to pursue new things. If it's a credible business and technology that's emerging, you can have smart people lean into it, and you can build expertise quickly. But overall, I'd say the similarities are stronger than the differences. Have an approach that is clear on what good looks like and go all-in behind it — say this is what the future is. Collaborate, take people with you, and don't use uncertainty about the future as an excuse for inaction.

With COP26 now behind us, how happy are you with the outcomes and the way forward from here?

COP26 came at a challenging time: nearly two years into the pandemic. It will be remembered by many as the business and finance COP, where the private sector mobilized in force behind net zero targets. The COP failed to deliver on some commitments such as finance for developing nations and more support for resilience and adaptation, but a real focus on countries' future climate action plans — and expectations that these become more ambitious within the next 12 months — will keep the pressure on national policymaking. The countries that do this well will future-proof their economies and see a big share of business innovation and investment.

Endnotes

1. Asian Development Bank, *Meeting Asia's Infrastructure Needs*. Manila: Asian Development Bank, 2017.
2. United Nations Climate Change, "Introduction to climate finance", https://unfccc.int/topics/climate-finance/the-big-picture/introduction-to-climate-finance.
3. United Nations Climate Change, "The Paris Agreement", https://unfccc.int/process-and-meetings/the-paris-agreement/the-paris-agreement.
4. G20, "G20 leaders' communique", Hangzhou Summit, 5 September 2016.
5. United Nations Environment Programme Inquiry and People's Bank of China, "Establishing China's green financial system", Report of the Green Financial Task Force, April 2015.
6. Otoritas Jasa Keuangan (OJK), "Roadmap keuangan berkelanjutan di Indonesia. (Roadmap for sustainable finance in Indonesia)", December 2014.
7. Bursa Malaysia, "Bursa Malaysia: CSR key to business sustainability", 5 September 2006.
8. Hong Kong Exchanges and Clearing, "How to prepare an ESG report", March 2020.
9. International Finance Corporation, Mercer, "Gaining ground — sustainable investment rising in emerging markets", March 2009.
10. Principles for Responsible Investment, "About the PRI", https://www.unpri.org/pri/about-the-pri.
11. The Asia Investor Group on Climate Change, "Overview", https://www.aigcc.net/overview/.
12. PwC, "Asset and wealth management revolution: The power to shape the future", March 2021.
13. HSBC, "HSBC sets out net zero ambition", 9 October 2020.
14. Mitsubishi UFJ Financial Group, "MUFG carbon neutrality declaration", 17 May 2021.
15. Global Sustainable Investment Alliance, "Global sustainable investment review 2020", July 2021.
16. Ibid.
17. Refinitiv, "Sustainable finance surges in popularity during H1 2021", 27 July 2021.
18. IFR (Refinitiv data), "Green bonds hit record before H1 ends", 25 June 2021.
19. Refinitiv, "Sustainable finance surges in popularity during H1 2021", 27 July 2021.

20. Prequin, "Future of alternatives 2025: Laying the foundations for infrastructure growth", 4 November 2020.
21. Macquarie, "Accelerating investment into sustainable infrastructure", 3 March 2021.
22. Macquarie, "Macquarie led consortium completes acquisition of the Green Investment Bank", 18 August 2017.
23. Macquarie, "How renewable energy infrastructure could accelerate Asia's green future", 23 September 2020.
24. Arabesque, "Are companies moving fast enough on reporting greenhouse gases", November 2020.
25. Ibid.
26. Bank Negara Malaysia, "Climate change and principle based taxonomy", Kuala Lumpur, April 2021.
27. Securities and Futures Commission, "Circular to management companies of SFC authorized unit trusts and mutual funds — ESG funds", Hong Kong, June 2021.
28. Diginex, "Who we are", https://www.diginex.com/.
29. RIMM, "Who are we?", https://www.rimm.io/home.
30. World Bank, "State and trends of carbon pricing 2021", Washington DC, May 2021.
31. South China Morning Post, "China's ETS a small but important step in climate change fight", 18 July 2021.
32. CDP, "Putting a price on carbon. State of internal carbon pricing by corporates globally", London, April 2021.
33. The Temasek Review, "Pathways to sustainability: Putting a price on carbon", https://www.temasekreview.com.sg/pathways-to-sustainability/putting-a-price-on-carbon.html.
34. SCMP, "Sustainability data start up Miotech plans to double asian headcount after bagging Guotai Junan, Sing's GIC as new investors", 25 August 2021.
35. Climate Action 100+, "Investor signatories", https://www.climateaction100.org/whos-involved/investors/.
36. Climate Action 100+, "Companies", https://www.climateaction100.org/whos-involved/companies/.
37. SCMP, "HSBC to end coal financing by 2040 after facing pressure from investors", 11 March 2021.
38. Reuters, "Japan's SMFG to halt all new financing of coal fired power plants", 12 May 2021.
39. Reuters, "Southeast Asia's largest bank DBS to phase out thermal coal financing", 16 April 2021.

40. MS&AD Insurance Group Holdings, "Initiatives to achieve net zero by 2050 (Part 2)", 25 June 2021.
41. Reuters, "S Korea's major insurers say will stop underwriting new coal power", 22 June 2021.
42. Finance Asia, "Transition financing heats up in Asia — part 1", 9 August 2021.
43. Asian Development Bank, "ADB backs first climate bond in Asia in landmark $225 million Philippines deal", Manila, 29 February 2016.
44. Asian Infrastructure Investment Bank, "AIIB, Clifford Capital establish financial platform to provide institutional capital access to infrastructure debt financing", 28 November 2019.
45. Hong Kong Monetary Authority, "HKMC Signs MoUs with five partner banks on infrastructure loans framework", 22 March 2021.
46. SCMP, "Budweiser APAC secures US$500m loan with interest rate tied to Brewer's ESG performance", 12 July 2021.
47. Leo Paper Group, "The first private company & manufacturer in Hong Kong to complete green financing for two consecutive years", September 2019.
48. Finance Asia, "Citi seals first green deal in Chinese dairy industry", 18 May 2021.
49. Bloomberg, "Energy transition's half trillion dollar year is even better than it looks", 21 January 2021.
50. Reuters, "Brookfield raises $7 billion for global energy transition fund", 28 July 2021.
51. BlackRock, "Temasek and BlackRock launch decarbonization investment partnership", 12 April 2021.
52. PwC, "The state of climate tech 2020", September 2020.
53. Ibid.
54. Ibid.
55. Dream Labs Innovation, "Investing in the improvement of mankind", https://www.dreamlabs.sg/.
56. Razer, "Razer establishes USD50 million 'Razer Green Fund' to support sustainability startups", 21 April 2021.
57. Conservation International, DBS Bank, National University of Singapore, Temasek, "The business case for natural climate solutions: Insights and opportunities for Southeast Asia", December 2020.
58. United Nations Environment Programme, "State of finance for nature", 7 May 2021.
59. Mirova, "Natural capital", https://www.mirova.com/en/invest/natural-capital.

60. The Meloy Fund, "Financing the transition to sustainable fisheries", https://www.meloyfund.com/.
61. Carbon Direct, "CO2 management", https://carbon-direct.com/advisory/.
62. Climate Investment Funds, "Climate finance: Investing in our planet", https://www.climateinvestmentfunds.org/.
63. United Nations Climate Change, "IFC China Utility-Based Energy Efficiency (CHUEE) program", https://unfccc.int/climate-action/momentum-for-change/activity-database/momentum-for-change-ifc-china-utility-based-energy-efficiency-chuee-program.
64. Green Climate Fund, "Project portfolio", https://www.greenclimate.fund/projects.
65. Green Climate Fund, "GCF spotlight: Small Islands Developing States (SIDs)", 1 July 2021.
66. Green Climate Fund, "FP090", https://www.greenclimate.fund/project/fp090.
67. Green Climate Fund, "FP066", https://www.greenclimate.fund/project/fp066.
68. Tropical Landscapes Finance Facility, "Tropical Landscapes Finance Facility", https://www.tlffindonesia.org/.
69. Alexandra Tracy, "Leading Asia's financial future — Hong Kong Green Investment Bank", European Chamber of Commerce in Hong Kong, October 2017.
70. Green Bank Network, "Malaysian Green Technology and Climate Change Centre", https://greenbanknetwork.org/malaysia-green-technology-and-climate-change-centre/.
71. Green Climate Fund, "FP153", https://www.greenclimate.fund/project/fp153.
72. Glasgow Financial Alliance for Net Zero, "Amount of finance committed to achieving 1.5°C now at scale needed to deliver the transition", 2 November 2021.
73. HSBC, https://www.hsbc.com.
74. Finews.Asia, "HSBC: Asia profits dip but expansion on track", 2 August 2021.
75. HSBC, "Shareholders back HSBC's net zero commitments", 28 May 2021.
76. HSBC.
77. HSBC Holdings plc, "Annual report and accounts 2020".
78. Citywire Asia, "HSBC PB raises $1.3bn in client funding for alts investments", 4 June 2020.

79. HSBC, "HSBC asset management hires climatech team", 25 May 2021.
80. HSBC, "HSBC launches Singapore's first Green Loan for SMEs", 16 March 2020.
81. International Finance Corporation, "HSBC, IFC 'real economy' Green Bond Fund raises $538M at final close for climate action", 18 March 2021.
82. HSBC, "How to drive investment in sustainable infrastructure", 29 September 2020.
83. Temasek, "HSBC and Temasek launch partnership to catalyse sustainable infrastructure projects in Asia", 30 September 2021.
84. Pollination, https://pollinationgroup.com.
85. World Resources Institute, "HSBC partners with WRI and WWF to scale next generation solutions to climate change", 20 May 2021.
86. HSBC, "Centre of Sustainable Finance", https://www.sustainablefinance.hsbc.com.
87. HSBC, "HSBC launches ESG solutions team", 27 July 2020.
88. Chartered Banker, "Chartered Banker Institute and HSBC UK join forces to support responsible and sustainable banking", 1 July 2021.
89. Temasek, https://temasek.com.sg/. Accessed on: 31 March 2021.

Chapter 2
INDUSTRY & WASTE

by Sandro Desideri

ANALYSIS

by Sandro Desideri

This chapter is dedicated to providing a concise overview of how decarbonizing the industry and the tightly related complexity of waste management is paramount to delivering a decisive contribution to achieving net zero GHG emission goals. It introduces available and upcoming technologies and considers the policies in place, their enforcement, and what further policies are required to support the 2050 Net Zero goals.

Industrial activities such as manufacturing and construction are recognized as being responsible for around one-third of global greenhouse gas (GHG) emissions. If we add the emissions associated with the generation of electricity and heat and cold for industrial uses, we obtain the highest level of emissions of any economic sector. Even without the indirect emissions of purchased heat and electricity, industrial processes contribute about a fifth of global emissions. Only three sectors (steel mills/metal smelting, cement kilns, chemicals and plastics) account for about 55% of industrial emissions, and the top 10 industries are responsible for about 90% of industrial emissions. Since 2020, Asia's GDP has outpaced the rest of the world (see Figure 2.1), and industrial emissions will become the main challenge for the continent.

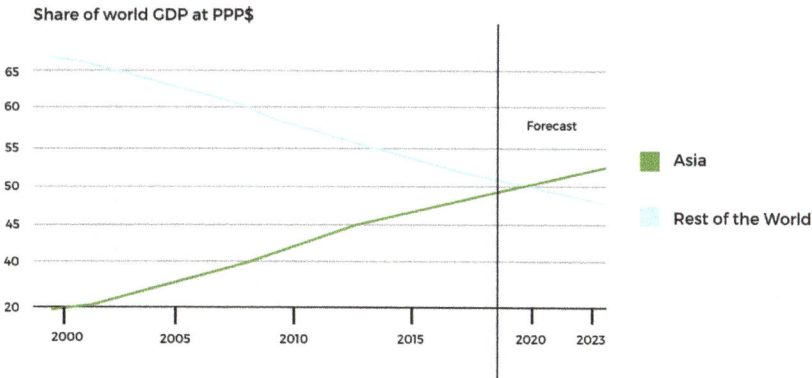

Figure 2.1. Asia's GDP trend.

Source: IMF 2020 @ valentinaromei.

59

Considering that most organic and solid waste is a by-product of industrial activity, waste management is an essential component of the net zero roadmap. The decomposing organic waste in landfills produces a gas that is composed primarily of methane, a GHG significantly responsible for climate change. Landfill gas can be recovered and utilized to generate electricity, fuel industries, and heat buildings. Beyond industrial by-products, global food loss and waste also annually generated 4.4 Gt of CO2 or about 8% of total anthropogenic GHG emissions. This means that food waste emissions are equivalent (87%) to global road transport emissions.

Asia-Pacific plays a major role in the creation of waste, producing 23% of the global output, the most of any region.[1] In East Asia, Hong Kong, Singapore, Taiwan, and China generate a combined 223 million tons annually. Of these four markets, Taiwan is seen as a world leader in recycling and zero waste, with high recycling rates and low waste disposal for its per capital income level. Taiwan's success with recycling is proof that the rest of East Asia can achieve a low-waste economy through effective policy, infrastructure, and education.

Definitions and Convergence to the Circular Economy Concept

Industry and Waste are two very contrasting subjects. The latest trends demonstrate that the future of mankind depends on how we manage the production of goods and generation of waste, as well as the consumption of goods and generation of garbage. A few core definitions will be paramount for the chapter.

Industry: Economic activity concerned with the processing of raw materials and manufacture of goods in factories.

Waste: Anything that is rejected as useless, worthless, or in excess of what is required, i.e., garbage, rubbish, or trash; however, we like this alternative definition: "A failure to take advantage of something".

The two are strongly related. Industry generates a product and Waste is the by-product of the process. Products generate more products and more waste in a spiral that filters down from the R2I (raw materials to industry) to B2B (business to business) and then B2C (business to consumer) segments. This is sometimes referred to as the "take, make, waste" process.

Circular Economy[2]: The process of reusing, sharing, repairing, refurbishing, remanufacturing, and recycling. Since the early 90s, Europe and most of the developed countries in the world (Japan, Singapore, Taiwan, and South Korea in Asia) have been concerned about environmental transition issues, and such concerns have been translated into directives and zero-waste programs that have become fundamental milestones in the decarbonization process.

The definition of a Circular Economy (CE) contrasts with today's prevailing concept of "linear economy", which supports the predominant structure of the value chain in industrial production, in which goods are produced, consumed, and discarded. In a CE, each product at the end stage of its life is considered as a resource that can be reused as a raw material. The CE is a cascade of options, rather than a single type of activity (such as "recycling" or "reuse"), minimizing the loss of value of the industrial process.[3]

The way an Industry utilizes and disposes of a product, as well as the way its Waste is managed, are two faces of the same coin. However, for the purpose of this book, we will treat the two items separately to better clarify the analysis of the subject.

Decarbonizing the industry is a core part of numerous government plans for reaching net zero targets and of many global industrial players.

In Asia, the Japanese government is leading the charge, becoming the first country to set a 2050 Net Zero goal. They have introduced and are actively driving the concept of a Green Hydrogen Economy.[4] Japan seeks to decarbonize its economy by leveraging hydrogen energy in transportation, industry, power production, and other fields, having formulated the world's first national hydrogen strategy in 2017. The Environment Innovation Strategy rolled out in January 2020 has included hydrogen amongst the fundamental technological innovations necessary to achieve global carbon neutrality by 2050.

The European Union (EU) is also partaking in the vision by implementing policies like the latest EU Energy Directive. The directive is supported by the introduction of "white certificates" (subsidies for implementing energy-efficient and energy-saving measures). The United Kingdom (UK),

South Korea, China, India, and the United States (US) are also significantly contributing to the decarbonization effort.

It is valuable to start by categorizing the Industry sectors:

- Primary sector (the raw materials industry)
- Secondary sector (manufacturing and construction)
- Tertiary sector (the "service industry")
- Quaternary sector (information services)
- Quinary sector (human services)

We will focus on the Primary and the Secondary sectors, considering the relevant carbon intensity:

- Heavy Industry (i.e., refineries, petrochemicals, steel mills, glass factories, paper mills, mining, cement, ceramics, metal smelting, etc.),
- Manufacturing Industry (i.e., machinery, electronics, semiconductors, etc.), and
- Transformation Industry (i.e., pharma, food and beverage, etc.).

Historical Development of Industry

Understanding the development of industrial activities in recent history is essential for a rational understanding of the current and future situation of decarbonization in the Industry sector, and the approach must be holistic to identify the causes and effects that are now parts of the custom of the manufacturing world. A good starting point is the analysis of the British industrial revolution in the second half of the 18th century. This development brought about profound transformations in the then-predominantly agricultural and largely rural societies in Europe and America and their consequent "reinventing" into industrialized urban societies. The English industrial revolution introduced machines and the power of steam. Much of the production activity and the goods that had once been carefully worked by hand (think of the textile and metalworking sectors) began to be produced in large quantities, generating great demand for raw materials and, consequently, the energy that was needed to sustain this development.

Fueled by the game-changing use of steam power, the Industrial Revolution began in Britain and spread to the rest of the world. By the 1830s, it reached the US. Modern historians often refer to this period as the First Industrial Revolution to set it apart from the second period of industrialization that took place from the late 19th to early 20th centuries, which saw rapid advances in the steel, electricity, and automobile industries. During the 19th century, Asia was marginal in global industrial developments. Only Japan stepped onto the path of fast industrialization at the end of the century. This situation is reflected in the CO_2 emissions level from industrial activities in Asia throughout most of the 19th and early 20th century — an absolute negligible amount compared to the US, UK, and Germany.

It was "Industry 1.0", and during the 18th century, mankind started to influence the environment by generating industrial pollution, poisoning the air, ground, and rivers. However, the world was conceived as being immense, with limitless and unexplored resources — there was a certainty that human activities would have no impact on the climate.

It is important to note that the evolution from Industry 1.0 to Industry 4.0 (see Table 2.1) is not a sequence of improvements that makes the previous version obsolete; rather, it is an upgrading of the industry toward greater efficiency, productivity, and less dependency on human labor.

Industry 1.0 is still here. And it remains a gigantic industry. Just the global oil and gas market alone is expected to grow from USD 4,677 billion in 2020 to USD 5,870 billion in 2021 at a CAGR of 25.5%, and to USD 7,425 billion in 2025 at a CAGR of 6%.[5]

Table 2.1.   The Four Industrial Revolutions.

INDUSTRY 1.0	INDUSTRY 2.0	INDUSTRY 3.0	INDUSTRY 4.0
Mechanization	Mass production	Automated production	Smart factories
Steam & water power	Assembly lines	IT systems & computers	IOT & autonomous systems
	Electrical power	Robotics	AI, machine learning

Current State of GHG Emissions in the Industry

According to the IEA's "Net zero by 2050" report,[6] the Industry sector is responsible for 30% of GHG emissions, and it is considered to be the least capable of aligning with the requirements of the net zero path. The report shows that the Industrial sector will represent 50% of GHG global emissions by 2035. Today, the sector represents 30% when compared with the three other sectors (Electricity, Transport, Buildings). In Asia-Pacific, the GHG emissions by 2040 will probably be 60% from industrial activities.

The reasons for such resistance are linked with the massive capital expenditure (Capex) involved in the heavy and mid industry and the relevant huge operational expenditure (Opex) that is generating jobs and returns to the service sector. We are talking about investments with a life-cycle of at least 20 years per project but much longer in terms of the overall mission and vision of each industrial corporation intimately linked to long-term R&D plans.

In the Asia-Pacific region, besides being a commercial challenge, the resistance to fully embracing a net zero strategy is also due to the high growth of populations, low-rate unemployment policies (resistance to technological innovation considered an occupational risk), and the status of being a "developing country". In many of the fastest-growing industrializing countries in the world, there is a lack of policies able to incentivize innovation. Furthermore, there is a lack of effective monitoring and enforcement regimes for industrialists not following environmentally friendly statutes.

By examining the portion of global GDP deriving from the categories of heavy industry and processing, we can obtain a clear assessment of their impact on GHG emissions. These industries are part of the process/manufacturing and construction sectors and are cumulatively responsible for 30% of global GHG emissions.

The Heavy Industry market accounts for around 10% of the global GDP. However, as of 2019, it emits about 22% of global GHG emissions: pure heat in the production process, making up 10% of global emissions. The steel industry alone was responsible for 7 to 9% of the global carbon dioxide emissions, inherently related to the main production process of reducing iron with coal.

The Manufacturing and Transformation sectors are a major part of the economy and account for nearly 16% of the global GDP in 2019, and are jointly responsible for around 8% of global GHG emissions.[7]

Unilever, one of the world's largest consumer goods companies that own famous brands in F&B, home care, and beauty and personal care, and with a footprint and sourcing in Asia, is a great example of a multinational company that pledged for net zero GHG emission by 2039. The plan encompasses specific measurable actions in sectors like the packaging of food, decarbonization of the energy for the processes, and overall efficiency manufacturing improvements — all actions that are driven by available technologies. Unilever's challenge is to make such technology implementations economically sustainable.

The Steps Ahead for Decarbonizing the Industry[8]

To obtain an alignment between the growth objectives of the global industry, the efforts for sustainability, and the achievement of the 2050 Net Zero target, we must consider some fundamental factors: the fragmentation of the sector into a large number of sub-categories with specific objectives and requirements, the need for a gradual approach, and the time factor, which does not give many opportunities for delay.

Generation and implementation of guidelines and regulations emerge as elements of absolute importance for obtaining a coordinated and sustainable fulfilment of the decarbonization objectives. However, the economic challenges and social impacts that will be generated by such implementations and, therefore, the obvious difficulties of the legislative work must be taken into account. In any case, it is absolutely clear that governments and regulatory institutions must lead the acceleration of coordination activities since it does not seem possible to obtain results without aligning the incentive policies and application of decarbonization. It has been estimated by Accenture that the application of a carbon tax for industry of about 40 USD/MT on industrial emissions could be an excellent starting point for Europe. In Asia-Pacific, it will be essential for governments to reach an agreement on how a carbon tax can be manageable and avoid impacts on the development capacity of some countries that do not enjoy the status of developed countries.

It is important to reflect on the fact that the financial cost of producing and placing on the market industrial products with a low impact in terms of GHG emissions will be substantially higher over a long period than the adoption of the current highly emission-intensive production systems. The duration of this period depends on the ability to globally scale the implementation of technologies, many of which already exist but are unable to be economically sustainable without reaching a large scale of implementation. It is clear that the gap generated during this period must be supported by the strict implementation of regulations and the provision of incentives in a measurable and monitorable manner. However, the reduction of industrial emissions to net zero is and will long remain a commercial challenge, influenced by factors such as: fuel prices, military conflicts, geopolitical situations, uncertainties in the adoption of new technologies deriving from a lack of standards, and the ability to cohesively group industrial activities in order to use shared infrastructures.

After a period of uncertainty, the markets will be able to identify the most appropriate paths for decarbonization. However, it can be predicted, especially in Asia, that governments will have to take on the burden of overcoming many market failures and barriers to entry that prevent the industry from securing the necessary investments to initiate the low-carbon transition.

It can be legitimately concluded that in order to ensure investments for the decarbonization of the industry, the following three levers are necessary:

- A widely implemented and accepted carbon pricing mechanism.
- An accessible funding mechanism to secure private sector investment.
- A fair and widely accepted policy reform to mitigate risk of "carbon leakage".

Future prices for carbon emissions and renewable electricity remain uncertain; however, Accenture[9] predicts the annual net value of industrial decarbonization will more than double in Europe between 2020 and 2030, from around 100 to just over 200 billion euros per year. The successful opening of China carbon trading in 2021 is a very important signal we are in the right direction and that Asia can become the biggest net value of industrial decarbonization any time between 2030 and 2040.

Trends and Technologies

Even though the Industry sector is very fragmented, we are comfortable saying that the most important decarbonization trends and technologies are around overall efficiency, the utilization of green hydrogen, waste heat recovery (WHR), carbon capture and storage (CCS), as well as biogas and biomass. The following are detailed breakdowns of each.

Efficiency refers to the overall industrial process. The majority of companies are focusing on energy efficiency since most transformative solutions, including electrification technologies and carbon capture, are not yet financially attractive. This process is the most urgent and easiest to achieve! One of the most effective ways to boost the efficiency of an industrial plant that makes use of electrical and thermal input (F&B, ceramics, paper mills, pharma, etc.) is to implement **combined heat and power (CHP)** systems for cogeneration (power and steam/heat) or trigeneration (power + steam/heat + cooling). Figure 2.2 outlines the combined heat and power process at a glance.

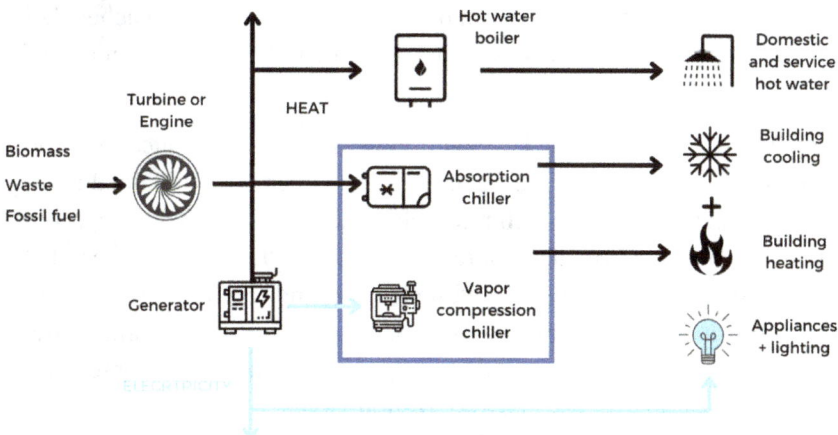

Figure 2.2. The CHP process.

Source: CO.DUE srl 2020.

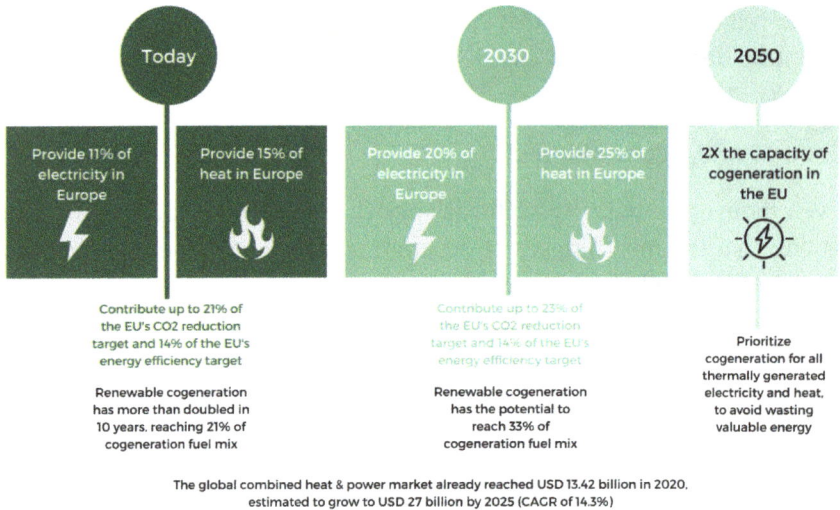

Today		2030		2050
Provide 11% of electricity in Europe	Provide 15% of heat in Europe	Provide 20% of electricity in Europe	Provide 25% of heat in Europe	2X the capacity of cogeneration in the EU

Contribute up to 21% of the EU's CO2 reduction target and 14% of the EU's energy efficiency target

Contribute up to 23% of the EU's CO2 reduction target and 14% of the EU's energy efficiency target

Prioritize cogeneration for all thermally generated electricity and heat, to avoid wasting valuable energy

Renewable cogeneration has more than doubled in 10 years, reaching 21% of cogeneration fuel mix

Renewable cogeneration has the potential to reach 33% of cogeneration fuel mix

The global combined heat & power market already reached USD 13.42 billion in 2020, estimated to grow to USD 27 billion by 2025 (CAGR of 14.3%)

Figure 2.3.   Europe's roadmap for CHP implementation 2020–2050.

Source: CO.DUE srl 2019.

The efficiency of the autogenerated power supply can reach 90% from an initial 30–40% in a non-combined solution. Figure 2.3 shows Europe's commitment to implement CHP fueled by natural gas and H2 + biogas as part of their net zero roadmap.

The global combined heat and power market already reached USD 13.42 billion in 2020, and is estimated to grow to USD 27 billion by 2025 (CAGR of 14.3%).[10]

Green hydrogen (H2) is a key solution for sustainable applications and a major driver toward the "hydrogen society". Green H2 is obtained mainly through electrolyzers fed with renewable energy and is, therefore, a zero-CO2 emission kind of fuel (more details in the "Energy" chapter of the book). Japan is leading the way with planning and regulatory frameworks, whilst Europe is leading the way in R&D and technology piloting. Green H2 can be an outstanding contributor to support industrial processes, making up at least 30% of the decarbonization target.

Green H2 can be used for the production of green ammonia (NH3), which will be fundamental for the decarbonization of fertilizers and the chemical industries. Green H2 in the form of hydro-methane (H2 blended with natural gas in 10% to 100% proportions) is becoming the preferred choice for all the new CHP applications involving gas engines and/or gas turbines.

According to Precedence Research, the green H2 market is estimated to reach USD 1.83 billion in 2021 and is projected to grow at a CAGR of 54% from 2020 to 2030.

Waste heat recovery (WHR) represents a huge market in industrial applications that are already getting valuable attention and incentives.

WHR Incentives Case Study (Italy)

White certificates

Also called Energy Efficiency Certificates, or Titoli Efficienza Energetica (TEE), white certificates are the main incentive mechanism for energy efficiency in the industrial, network infrastructure, and services and transport sectors. One certificate is equivalent to saving one Ton of Oil Equivalent (TOE). The Gestore dei Servizi Energetici (GSE) is the Italian institution that recognizes a certificate for each TOE of savings achieved, thanks to the implementation of the TEE. To this end, all the subjects admitted to the mechanism are included in Gestore Mercati Elettrici's (GME) Electronic Register of Energy Efficiency Certificates. The economic value of the securities is defined in the trading sessions on the market.

Asia does not implement policies like WHR certificates yet, even in the more developed countries like Japan. There is a set of different and very peculiar policies that require alignment and concurrence to make the journey to net zero effective.

Several technologies can still be used to recover waste heat — the Organic Rankine Cycle (ORC)[11] is a mature and effective technology that is still improving, thanks to new materials, innovative manufacturing solutions, and implementation of sophisticated control systems.

The ORC technology is vastly applied in refineries, steel mills, and glass factories, and in many industrial processes that are producing a large amount of heat. It is also the most effective technology in use for geothermal plants. Figure 2.4 shows the basic principle of the ORC technology.

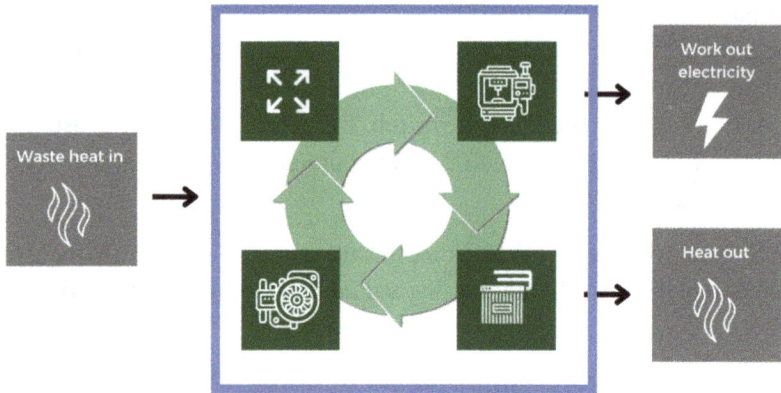

Figure 2.4.　Organic Rankine Cycle.
Source: Orcan International, 2021.

According to Grand View Research, the global waste heat recovery system market size was valued at USD 54.3 billion in 2019 and is expected to grow at a CAGR of 8.8% from 2020 to 2027.

Hybridization & Renewables maximization is a very important and impactful solution for the transition to a fully decarbonized industry. This solution decreases the impact of fossil fuels in the generation of electricity by integrating renewables and energy storage capabilities in the energy mix. Its advantage lies in delivering a fit-for-purpose solution that overcomes the big problem created by the intermittency of the renewable energy supply. Figure 2.5 shows a classic schematic for a Renewables + Gas Genset + BESS (Battery Energy Storage System) Hybrid Power System.

Hybrid generation plants are becoming more widespread among energy-intensive industries focused on reaching net zero GHG emissions. These industries have decided to invest in renewable resources to take advantage of government concessions and decrease their dependence on fossil fuels. However, the biggest concern in the industrial process is the categorical imperative to supply energy continuously. Renewables like solar and wind energy suffer from an intermittence problem, meaning that a cloudy day could significantly disrupt the production process.

Hybrid applications are a potential solution. Examples available in Asia are: (1) cement factories that use lime excavation areas to implement solar plants to charge batteries and use biomass pellets for the kiln

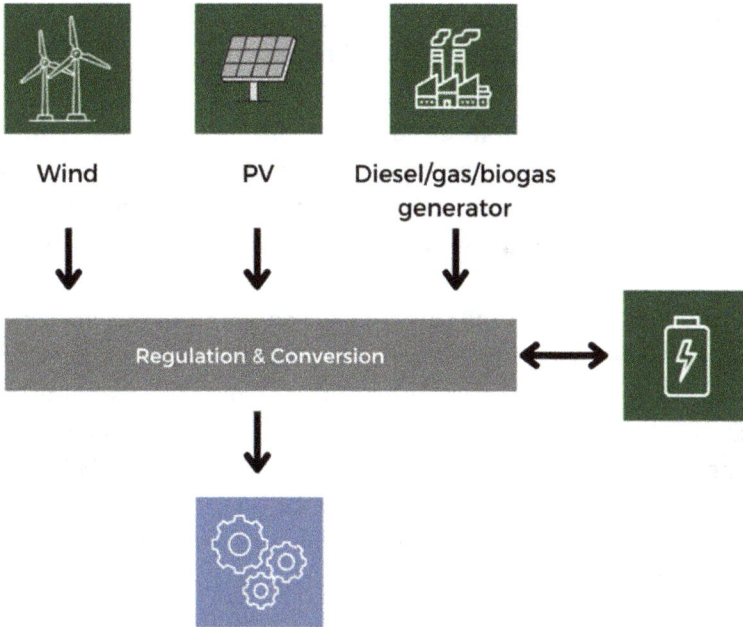

Figure 2.5. Hybrid Power System, which combines multiple sources to deliver non intermittent electric power.

Source: Link4Success Ltd, 2021.

process, and (2) paper mills that use solar/wind for water demineralization and to charge batteries that energize the plant. Biogas is also used for the generation of steam and management of electrical peaks by using combining conventional heat and power systems. Similar applications are available for most industrial processes.

The next step toward net zero and workable economics would be to replace diesel engines with gas engines, using mini LNG and/or Biogas as fuel. This solution will be able to fully satisfy the requirement of managing the supply-demand of intermittence and peaking electricity. The global hybrid power solutions market was valued at USD 1,352.4 million by 2025 and is estimated to denote a CAGR of 9.5% from 2017 to 2025.[12]

Biogas & Biomass can provide a source of carbon-neutral baseload power, when combined with carbon capture, utilization, and storage (CCUS) applications, would be able to reach negative CO_2 emissions considering that biomass is carbon neutral and that CCUS is already taking neutral CO2 out from the emissions. Biomass comes from crops and sustainable forestry that absorbs CO2 from the atmosphere, keeping the whole balance unaltered. Furthermore, an additional impact on GHG generation must come from the development of sustainable Biomass. Transforming feedstock into Biomass & Biogas can reduce methane (CH4) emissions originating in manure, unmanaged forestry, and waste from agriculture. Importantly, methane's GHG impact is 14 times higher than that of CO2.

It must be emphasized that this section primarily focuses on Sustainable Biomass. Biomass derived from unsustainable agricultural practices and plantations, as well as from deforestation, is not a viable choice for decarbonization. On the contrary, it can be its worst enemy.

One interesting application is the implementation of CHP systems based on biogas for slaughterhouses. The biogas plant is operated with selected fractions of the pig slaughtering process such as pig blood, minced hindgut, as well as contents and fat from dissolved air flotation. The biogas is generated using anaerobic fermentation tanks or pools, depending on the size of the operation. The great advantage of such applications is that they offer more than just an automated producer of electricity and thermal power. They also reduce GHG emissions while addressing the issue of disposing difficult-to-process materials.

China currently has about 5,000 pig slaughterhouses. Shuanghui, the largest player and an affiliate of Smithfield Foods, slaughtered around 15 million pigs in 2020. The implementation of Biogas & Biomass technologies is rising and is already implemented in major slaughterhouses across the country.

Another interesting application of biogas generation is used in Thailand, a country with a substantial record of using biogas systems for the industry. The Thai Association for the Development of Biomass and Biogas is the most powerful in Souteast Asia. Take Thai San Miguel Liquor (TSML) distillery in Bangkok for example. The system implemented is able to mitigate GHG emissions caused by the decomposition of wastewater from the distillery, converting such emissions into biogas that then feeds the gas engines. The electricity is then used to replace the fossil fuel energy, which powers the TSML boilers and is also exported to the Thai national grid. The process reduces approximately 87,000 tonnes of emissions per year.

A notable example of sustainable biomass applications in industrial decarbonization comes from Japan, where the "Net zero by 2050" pledge is imposing restrictions on the use of coal in industrial applications. The Ministry of Economy, Trade and Industry has imposed strong pressure on all companies that work in energy-intensive sectors and emit GHGs, e.g., cement kilns, steel mills, refineries, etc. Before 2030, all plants that use coal must become more than 43% efficient or shut down. This is significant, because the average efficiency of a 10-year-old coal plant is between 34% and 37% and the cost of upgrading or replacing such a plant exceeds any reasonable economic consideration. As a result, replacing coal with sustainable biomass has become one of the best and most convenient solutions. Coal is blended with biomass in the form of black pellets and can sustainably bring the plants to their end-of-life cycle by reducing CO_2 emissions equal to the percentage of blending.

The technology that allows and controls the efficient blending of sustainable biomass is already available and consolidated. Other elements to consider as an important complement to the technology are the emission control systems (consolidated technology), Carbon Capture systems, and digitalized management of energy efficiency.

Waste

First and foremost, effectively managing waste can drive positive and needed changes in behavior and provide an occasion to promote circular economies. Most importantly, it highlights the need for a massive urban and industrial infrastructure upgrade. Most of the cities and industrial parks in Asia have been developed without any consideration for sustainability. This can change.

As introduced in the opening of this section, waste is a subject that involves many different factors. The collection, segregation, and recycling of waste are relevant to any country's infrastructure. There is an ongoing conversation in every municipal office in Asia about the trade-offs of very cheap incineration landfills versus very expensive sustainable treatments.

Ultimately, what are the suitable solutions in the waste-to-energy pipeline in a regulated environment?

Waste is a subject with considerable social impacts.

Specifically, in Asia, we have to consider the following:

- According to the UN Environment Programme,[13] urban food waste is predicted to increase by 44% from 2020 to 2050.
- 40% of the world's waste ends up in open dumps.
- 3.5 billion people in the world lack access to proper waste management.
- Technologies to fix waste are generally affordable for developed countries but not for developing countries.

Evolution of Waste Management

As a significant example linked to industrial development, one of the first organized solid waste management appeared in London at the end of the 18th century, where a waste collection and resource recovery system was established around the "dust-yards".[14]

Coal ash ("dust") was the main constituent of municipal waste. Such waste had market value for brick-making and as a soil improver. Encouraged by the profitability of recovering such waste, dust-contractors started to recover 100% of the residual waste remaining after readily selling items. This was an early example of organized municipal waste management. The dust-yard system worked successfully up to the mid-1850s, when the market value of "dust" collapsed. However, the experience was fundamental in facilitating the relatively smooth transition to an institutionalized solid waste management system in England.

Until the mid-19th century, the concerns about waste management were fundamentally relevant to big communities' garbage and sewage production, while agricultural waste was well managed (with excellent results throughout China). There was a small amount of "industrial waste", mainly due to the inefficient production systems of ceramics, pots, tools, and weapons. Waste from construction was almost non-existent due to the difficulties of transportation, forcing producers to use whatever was available.[15]

Today's Challenges of Waste

Collection, segregation, and recycling are the future of more harmonious humanity but are not viable solutions for developing countries, representing 70% of the world's population.

Landfills are not a sustainable solution in Asia or any densely populated area, as they cause odors, sanitary problems, and wasted land. The biggest challenge now is getting rid of landfills and sanitizing them. Besides a granular implementation of the circular economy principle and methods, there is an evident need for technology to help alleviate the problem.

One technology of interest and with a potential huge impact is Waste to Energy (WtE), also known as Energy from Waste (EfW), a highly exclusive and sought-after sector because of its renewable nature and indispensable role in the implementation of a Circular Economy, and, of course, its high returns from operations.

According to Veolia, a French multinational and a global leader in environmental infrastructure, the global WtE market is projected to reach USD 50 billion by 2027, growing at a 4.6% CAGR in the 2020–2027 period.

Waste-to-energy[16] is the process by which the energy contents of municipal solid waste (MSW) are converted into heat or electricity using various technologies. The major types of waste-to-energy technologies include landfill gas utilization, anaerobic digestion, and thermal treatment with energy recovery.

Apart from generating power, waste-to-energy provides an alternative way for the disposal of MSW, other than using a landfill. As land is scarce in many Asian countries (i.e., Hong Kong, Japan, China, etc.), developing more large-scale landfills is not an option. Thermal treatment is a very suitable application of the waste-to-energy process. Under controlled conditions, heat can be used to extract energy from organic waste.[17] The primary function of thermal treatment is to reduce the volume of MSW, with the recovered energy becoming a by-product of the treatment process.

Thermal treatment systems can be an integral part of an integrated MSW management facility, where a recycling facility serves as a pre-treatment system. The recycling facility can increase the material recovery percentage and also enhance the overall efficiency of the thermal treatment process. In a thermal treatment system with energy recovery, MSWs are used to produce heat and gas using combustion and gasification.

As the population increases, waste is piling up faster and faster across the Asia-Pacific region. According to Mordor Intelligence, by 2050, Asia-Pacific is expected to have around 1.3 billion tonnes of waste. Moreover, the daily per capita waste generation in the region is expected to increase by nearly 40% by 2050. Thus, increasing municipal waste volumes and demand for energy are expected to create even more opportunities for the waste-to-energy market in the near future.

China operates the largest waste-to-energy capacity of any country, with more than 300 plants in operation. This capacity has increased annually by 26% over the past five years, compared with just an average growth of 4% in OECD countries. Annually, China produces over 220 million tonnes of municipal waste per year.

At least 15 of the most technologically advanced MSW facilities in China have been developed and managed by Veolia, whose recent acquisition of Suez and the confirmation of its "ecological transformation" strategy are setting up the company to become an important player in the MWS sector across the Asia-Pacific region. It can generate convergence between investors, public-private partnership (PPP) models, and innovative technologies.

The waste-to-energy plant of Laogang, close to Shanghai, can generate up to 1.5 billion kilowatt-hours of power while the slag from the burned waste can be recycled into building materials. Shanghai, the largest city in China with more than 26 million people, generates over 20,000 tonnes of waste per day. This represents about one-third of the volume that can be processed at Laogang. Wu Yuefeng, the plant's chief engineer, says, "If you dump all of the garbage generated by the residents of Shanghai in one day into the Hongkou Football Stadium, it would pile up into a 21-meter-high hill, but after treatment, we can reduce this to only 2% of its original weight and 1% of its volume." The incineration site is part of the Laogang solid waste complex, which covers 29.5 km^2. This makes it the biggest solid waste treatment base in Asia, and it has processed over 75 million tonnes of material since it became operational in 1989.

In Singapore, the National Environmental Agency (NEA) and Public Utilities Board, Singapore's National Water Agency, announced the construction of the world's first integrated waste and water treatment facility — Tuas Nexus, which will be energy self-sufficient by harnessing synergies from the Tuas Water Reclamation Plant (WRP) and Integrated Waste Management Facility (IWMF). This is expected to result in carbon savings of more than 200,000 tonnes of CO_2 annually, equivalent to taking 42,500 cars off Singapore's roads. In addition, compared to building the two as standalone facilities, integrating both will result in land savings of up to 2.6 hectares — about the size of four football fields.

The Tuas WRP, a key component of the PUB's Deep Tunnel Sewerage System (DTSS) Phase 2 project, will have an initial treatment capacity of 800,000 m^3 per day, the equivalent of 320 Olympic-sized swimming pools. Unlike conventional water treatment plants, it can treat both domestic and industrial used water streams from two separate deep tunnels. In a first for Singapore, the Tuas WRP will also be capable of treating industrial water that can be recycled back into the plants.

Besides incorporating advanced physical, biological, and chemical treatment processes, the Tuas WRP will also house the largest membrane bioreactor facility in the world. This will enable the plant to be more energy-efficient while occupying less space compared to existing water treatment plants. A NEWater factory to be built on the rooftop of the Tuas WRP will boost the PUB's NEWater production capability, further ensuring a robust and resilient supply of water for Singapore that can meet its growing water demand and help it cope with the impact of climate change.

Call for action: The future of sustainable Industry & Waste management lies in public and private sector collaboration[18,19]

Collaboration between the public and private sectors is a success factor for successfully decarbonizing the industrial sector. A variety of actions and technologies that the private sector should consider on the journey toward decarbonization have been reported in this section. However, to obtain successful outcomes, the public sector's role is also critical.

Without robust action from governments, industrial companies will be at a competitive risk, given the need for investments and the uncertainty of the development of technological innovation. It is of the utmost importance for the public and private sectors to be in sync during this very complex transition. As a matter of the fact, both industrialists and corporates need predictability on costs in order to avoid extra costs during the transition and prevent "carbon leakage", in which industries lose business because the energy transition undermines their competitiveness in the international market.

In order to avoid succumbing to competitive risks, industrial companies need vigorous support from governments, taking into consideration their exposure to large capital and operational expenditure while under the uncertainty of the results of the effective implementation of technological innovation. The public and private sectors have to be in sync in order to provide industrialists with a decarbonization framework, with a predictability on transition costs and the prevention of "carbon leakage".

The public sector can help accelerate industrial decarbonization through the following actions:

- Providing a framework that will support companies to internalize carbon's hidden cost.
- Designing policies that avoid penalizing first-movers.
- Setting a carbon price mechanism with a high base price, increasing predictably over time as a guide to technological innovation and investment.
- Exploring a carbon-labeling standard for products.
- Introducing a carbon tax to compensate against competing imports and thereby preventing carbon leakage.
- Implementing quotas to increase the use of low-carbon products in construction and refinery products, supporting the scale-up of recycling.
- Stimulating the hydrogen economy on both the supply and demand side through a broad set of measures, including quotas and tax breaks.
- Consolidating and integrating international and national funding mechanisms into a streamlined single application process.

We believe that private and public sectors can deliver an accelerated and value-generating industrial decarbonization by working collaboratively.

The future of waste is uncertain because waste management remains a challenge with several complications. Such complexities are very evident in developing countries like Indonesia, Bangladesh, and other countries in Southeast Asia. There are plenty of factors impacting the social stability of populations in such countries, considering that open dumps are unfortunately still a source of income for the poorest in the world.[20]

Relevant solutions rely on a very similar approach to the one defined for the future of sustainable industrial development: partnership! The biggest difference is that waste management partnerships cannot only be between the private and public sectors, but between developed and developing countries as well. Waste management is an important tool for curbing climate change, and such methods should not remain in the hands of those who have nothing to gain, nothing to lose, and (mainly) nothing to invest.

STARTUP CASE STUDY
Wildfire Energy

by Sandro Desideri

An interesting technology based solution is offered by Wildfire Energy (WE),[21] an Australian startup based in Bribane (Queensland) that is working on getting energy out of landfills.

Landfills are expected to be drastically reduced in number and size in the near future, in consideration of its potential dangerous effects on human health and the environment. The generation of leachate can contaminate groundwater, and methane is produced, which is a potent GHG. In addition, where recyclable waste is landfilled, materials are unnecessarily lost to the Circular Economy or energy generation.

Starting from Europe and the US, almost every country is now implementing restrictions in order to reduce the amount of biodegradable waste and to secure the collection of the methane generated.

The problem that WE is addressing now is relevant across three specific points:

1. Non-recyclable solid waste is being landfilled at an ever-increasing rate, causing pollution and contributing 5% of global carbon emissions.
2. Incineration or waste-to-energy technology is not suitable for smaller-scale distributed projects or for the production of hydrogen.
3. The transition to net zero will require huge amounts of renewable electricity and green hydrogen to replace existing fossil fuels.

The solution — Moving Injection Horizontal Gasification (MIHG) — proposed by WE is aimed at supporting the generation of syngas out of the landfill in combination with methane. Such syngas can be utilized for power generation using reciprocating gas engines and, being a form of renewable energy, can be used to supply power to an electrolyzer's station to generate green hydrogen.

The MIHG technology is suitable for landfills, with a rate of filling less than 100,000 tons per annum of waste.

The claimed benefits of MIHG are the following:

Carbon negative — Delivers negative net GHG emissions when processing general wastes destined for a landfill.
Renewable energy — Produces renewable energy products such as electricity, hydrogen, syngas, and biofuels/chemicals.
Reduced landfilling — Reduces the quantity of waste sent to a landfill.
Circular economy — Can be integrated with material recovery facilities to increase waste recycling and transition to a more circular economy.

The business model implemented by WE is very efficient: (1) The waste supply agreement with the local landfill authority provides an income in the form of a "waste gate fee" (USD/t of waste), and (2) the establishment of a special purpose vehicle supported by a project investor generates revenue from the sale of electricity via a power purchase agreement with the local electrical utility (eventually incentivized by a feed-in-tariff) and the sale of green hydrogen for transportation use or storage. It has been calculated that the cost of the green hydrogen generated would be below 2 USD/kg and therefore be competitive.

The payback to investors could be in the range of 2 to 6 years while the lifecycle of the system is more than 15 years.

According to WE, the total accessible waste disposal market has a value of more than USD 250 billion per annum, with around 100,000 projects initiated every year. WE believes that around 10,000 projects are aligned with the characteristics of the technology and therefore can expect a service addressable market of around USD 25 billion by 2028. Beside the waste disposal market, WE's technology is also addressing the USD 500 billion green hydrogen market.

WE received several grants and funding to date: USD 1 million for the conversion of agricultural residues into hydrogen from CSIRO, and USD 0.4 million from the Queensland government for the commercialization of a MHIG module (Proton project).

The CEO of WE, Greg Perkins, a former Shell gasification and hydrogen unit manager, has indentified 2022 to be the year for further seeding while Series A is estimated for 2023.

WE's 2021 FY income was around USD 300,000, and its existing customers are the BMI Group (waste supplier for the proton project), Pure Hydrogen (hydrogen off-taker from the Proton project), Hitachi (cooperative development for processing food waste in Japan), and SIMS (a global recycling company).

All of WE's commercial projects implemented to date are relevant modular conversion in energy of 5,000 tons per annum to 10,000 tons per annum of waste (capacity up to 1 MWe) or conversion in green H2 (200 to 400 tons per annum).

WE comprises an all-Australian team consisting of specialists in the sectors of gasification processes, power generation, and project management, and the team is experienced in municipal BOO (build-own-operate) and PPP projects.

The impact of such technology can be massive for all minor landfills that are not attracting attention. Southeast Asia has thousands of such landfills, mainly around major metropolitan areas.

CORPORATE CASE STUDY
Aditya Birla Group — Ultratech Cement

by Sandro Desideri

UltraTech Cement Limited[22] is part of the Aditya Birla Group, one of the biggest conglomerates in India, with activities in metals, pulp and fibers, textiles, telecom, cement, and renewable energy. UltraTech is also the largest cement company in India, with a capacity of 116.8 million tonnes per annum (MTPA). Outside of China, UltraTech is the only company to produce over 100 MTPA of cement in a single country. In addition to India, the company also operates in the United Arab Emirates (UAE), Bahrain, and Sri Lanka.

Cement production is a very carbon-intensive manufacturing process, starting with the collection of raw material (limestone, clay, and marl extracted from quarries by ripping or blasting using heavy machinery), crushing and grinding the raw material, blending the raw materials in the right proportions, burning the mix in a kiln, grinding the burned product (known as clinker), and finally packaging the product.

The manufacturing of cement produces about 0.9 kg of CO_2 for every kilogram of cement. Every year, over 4 billion tonnes of cement are produced, and this activity is responsible for around 8% of global CO_2 emissions.

UltraTech is committed to the mitigation of climate change through a holistic approach to decarbonizing its operations. It has 23 integrated cement manufacturing units, 27 grinding units, and one clinkerization unit. Each of these units has implemented various initiatives such as digitalization, alternative fuel, and materials usage, and the adoption of renewable energy sources to decarbonize their operations. Carbon performance is one of the key pillars of the company's sustainability strategy.

For this reason, Ultratech also remains committed to transparent disclosure. Since 2013, it has been measuring its carbon footprint and disclosing it through the Carbon Disclosure Project's (CDP) platform. (The

CDP is an international non-profit organization based in the US, Germany, India, China, Japan, and UK, which supports companies in disclosing their environmental impact.)

The most efficient way to obtain sustainable results is to commit to validated and certified targets, and following this path, Ultratech decided to join the Science Based Targets initiative (SBTi), a partnership between the CDP, United Nations Global Compact, World Resources Institute (WRI), and the World Wide Fund for Nature (WWF) and has been designed to drive ambitious climate action in the private sector by enabling organizations to set science-based emissions reduction targets.

Ultratech has committed to reducing GHG emissions from clinker blending by 27% per ton of cementitious material by FY2032 from an FY2017 base year. It has also committed to reducing GHG emissions from indirect emissions through the use of power, heat, or steam supplied by external parties by 69% per ton of cementitious material within the same time frame. The target boundary includes biogenic emissions and removals from bioenergy feedstocks.

To understand the scale and importance of Ultratech in India's cement market, the following numbers need to be considered: (1) India marketed 321 million tons of cement in 2021, of which Ultratech — a USD 6 billion revenue company — dominates with 31% of the market share, equivalent to 1.7 billion of bags of cement, and (2) the captive power capacity for the production is 1,170 MW, of which 273 MW are from green power generation (waste-to-energy, solar energy, windmills). Ultratech is, by far, the biggest producer of cement in India; its major competitors are Shree Cement (37.9 MTPA 2021), Jaypee Cement (33.8 MTPA 2021), and ACC (33.4 MTPA 2021).

INTERVIEW
Gary Juffa
Governor, Papua New Guinea

by Sandro Desideri

We had the pleasure of interviewing the Hon. Governor of Oro Province, Gary Juffa, on the subject of the opportunities and challenges that decarbonization is presenting to Papua New Guinea (PNG).

PNG is rich in natural resources, including minerals, oil, gas, timber, and fish, and cash crops such as coffee, palm oil, cocoa, copra, rubber, tea, and spices, which contribute significantly to PNG's overall development. Several mining, oil, and gas companies are currently operating in PNG, and the operations of these companies have generated significant wealth to PNG's economy. Landowners affected by these developments also receive royalties from those operations. However, this wealth has not been fully translated into tangible social and economic development across the country, which is now facing the great challenge of directing the sustainable income from the exploitation of natural resources to the development of an environmentally friendly and decarbonized industry and economy. PNG represents an important laboratory on how developing countries can be beneficiaries of decarbonization-related investments and contributors to the fight against climate change.

Gary Juffa is a PNG politician and Member of the 10th Parliament of PNG. He founded the People's Movement for Change party. He was first elected to the 9th Parliament of PNG as the Governor of Oro Province (also known as Northern Province) in July 2012.

Juffa is an outspoken opponent of logging activity in his home province and across the country. In March 2018, he led the shutdown of controversial logging operations at Collingwood Bay, seizing equipment and arresting allegedly illegal workers. In Parliament, he has repeatedly called for action against illegal logging on customary land. He is recognized as a prominent leader in the modernization of the country while defending the incredible wealth of the PNG ecosystem.

Considering the subject, what is the actual situation of PNG and the industrial developments in the country relevant to the decarbonization aspect? What are the major challenges and opportunities?

The major challenges are posed by the deforestation of PNG perpetrated by the many companies from Southeast Asia that are actively logging the timber from virgin forest areas with the goal of implementing plantations of cash crops. Quite often, deforestation is only targeted for the trading of very valuable timber, and the plantations are not developed and left in their initial states. This is a big risk for our ecosystem, which has 7% of the world's biodiversity. The opportunities are in the development of environmentally friendly and net zero emission crops process industry and in the sustainable exploitation of forestry products.

As a political leader of a developing country, what do you see as your role in the decarbonization process of PNG?

The depletion of our ecosystem and the lack of opportunities to develop it into a sustainable resource for our country is the threat I am fighting against every single day in parliament. I am doing so by cooperating with organizations such as CIFOR-ICRAF,[23] which is doing a terrific job in PNG by supporting and enforcing policies against any illegal logging activities. These organizations are driving a collective educational effort about the value of being part of a big decarbonization strategy. As a matter of fact, the goal is to preserve the value of our land as a fundamental resource for the world and, in the longer term, to enjoy revenues from Carbon Credits generated by our policies and developments. I would like to remark that it is very important to reward the "good guys" in our country, I mean, the part of the population that is devoted to the protection of the natural wealth of PNG. I am fully motivated to work very hard in this direction.

What are the partners you would like to have in the process of decarbonizing the industry in PNG?

As mentioned before, we are enjoying the support of organizations like CIFOR-ICRAF, and also of the Asia Development Bank, the Australia Department of Foreign Affair and Trading, and many others. A special mention is the European Union's (EU) genuine interest in supporting the preservation of our ecosystem. However, we also need progressive corporates with a clear vision of their role in the net zero strategy that would be able to help us to establish and grow the processing industry of our resources, such as Vanilla, Cocoa, Coffee, etc. We are considered a true greenfield in such sectors, but we saw this as a great opportunity to do well and in a truly net-zero-oriented way from the start. The creation of related jobs will help grow the overall sentiment and capabilities of the nation.

What are the technologies and innovation you are considering as the most attractive for industry decarbonization in PNG?

We are very open to implementing technologies aligned with our overall net zero GHG emissions goal. Energy is key for the sustainability of our industrial development; therefore, renewable energy projects are paramount, and we are already implementing hydropower projects. However, we are also looking into each and every technology that can further help with a viable and green electrification of the country: smart grids and microgrids, adoption of energy efficiency systems, combined cooling, heat, and power out of our natural gas resources, and biomass generation out of sustainable forestry and crops waste.

What do you think would be the role of private investors in the decarbonization of industrial activities in PNG?

It is fundamental for PNG to attract more private investors. However, we need investors and corporates with a genuine plan to support the sustainability of their operations in PNG. We had quite a number of bad experiences with mining companies and their operations in the past. I am personally very keen to avoid this from happening again.

What is your model country or company for sustainable development?

During our mutual introduction, we discussed the contents of your book, and you mentioned a few multinational companies that are striving for net zero. Considering the industrial sustainable development roadmap, which I mentioned earlier, to be optimal for PNG, I am quite captivated by the Unilever model since it involves a deep understanding of how the processing industry must cope with the decarbonization requirements. We need to consider which other companies would be interested to invest in PNG, according to our available resources, in what would be a long but productive journey for the whole world.

How would you describe your experience at COP26, and how would you rate the impact of the international cooperation on the achievement of the decarbonization target?

I found the EU and other parties very receptive to the importance of supporting our efforts in defending PNG's rainforest ecosystem. We understand that there are many funds potentially available to support us, but my dilemma is knowing how to access these funds and making the whole process a success.

What do you expect to be the social impact of decarbonization activities in PNG?

As per my previous answer, I am expecting that the investments related to decarbonization will drive job creation, social stability, and overall educational advancement in my country.

Endnotes

1. Oliver Ironside, "How well does Asia recycle", *Earth.Org*, 2020, https://earth.org/asia-recycling/.
2. Catherine Weetman, *A Circular Economy Handbook for Business and Supply Chains: Repair, Remake, Redesign, Rethink*. London, 2016.
3. Ellen MacArthur Foundation, "Towards a circular economy: Business rationale for an accelerated transition", 2015, https://ellenmacarthurfoundation.org/towards-a-circular-economy-business-rationale-for-an-accelerated-transition.
4. Japan Ministry of Economy, Trade and Industry, "Japan's roadmap to 'Beyond zero carbon'", 2021.
5. Businesswire, "Global oil and gas market report 2021: COVID-19 impact and recovery — forecast to 2025 & 2030", 3 March 2021, https://www.businesswire.com/news/home/20210303005405/en/Global-Oil-and-Gas-Market-Report-2021-COVID-19-Impact-and-Recovery—Forecast-to-2025-2030—ResearchAndMarkets.com.
6. International Energy Agency, "Net zero by 2050: A roadmap for the global energy sector", 2016, www.iea.org.
7. Businesswire, "Global carbon footprint management market (2021 to 2027) — by component, deployment mode, type, end-user industry and region", *Businesswire*, 2019, https://www.businesswire.com/news/home/20220308005855/en/Global-Carbon-Footprint-Management-Market-2021-to-2027—by-Component-Deployment-Mode-Type-End-user-Industry-and-Region—ResearchAndMarkets.com.
8. Link4Success Ltd, www.link4success.com.
9. Accenture, "Energising industry. Generating > 200 billion per year by 2030 through European industrial decarbonization", June 2021, https://www.accenture.com/_acnmedia/PDF-147/Accenture-Energizing-Industry-Through-European-Industrial-Decarbonization-Full-Report.pdf.
10. Businesswire.com, "$24.12 billion opportunities in the automation in combined heat and power global market, 2021–2030 by component, control and safety system", *Businesswire*, 2022, https://www.businesswire.com/news/home/20220330005627/en/24.12-Billion-Opportunities-in-the-Automation-in-Combined-Heat-and-Power-Global-Market-2021-2030-by-Component-Control-and-Safety-System—ResearchAndMarkets.com.
11. Orcan International Ltd, http://iam-tv.org/home.html.
12. Markets and Markets, "Hybrid power solutions market by system type (solar-diesel, wind-diesel, solar-wind-diesel), power rating (up to 10 kW, 11 kW–100

kW, and above 100 kW), end-user (residential, commercial, telecom), and region — global forecast to 2021", *Markets and Markets*, 2021, https://www. marketsandmarkets.com/Market-Reports/hybrid-power-solution-market-121425179.html.

13. United Nation Environment Programme, https://www.unep.org.
14. Costas A. Velis, David C. Wilson and Christopher R. Cheeseman, "19th century London dust-yards: A case study in closed-loop resource efficiency", *Waste Management*, **29** (2009) 1282–1290.
15. Matthew Gandy, *Recycling and the Politics of Urban Waste*. Earthscan, 1994.
16. SkyRenewable Pty Ltd, "Energy from waste outlook", 2020–2021.
17. HK RE Net, "Thermal treatment", *HK RE Net*, https://re.emsd.gov.hk/ english/energy/thermal/ther_tec.html.
18. UK Government, "Industrial decarbonization strategy", March 2021.
19. Accenture, "Energising industry".
20. Zoe Lenkiewicz, "Peeping into the future of waste", 10 December 2015, https://zlcomms.co.uk/silver-bullets-or-stepping-stones-the-future-of-waste/.
21. Wildfire Energy, "Wildfire Energy: A revolution in energy from waste", https://www.wildfireenergy.com.au/.
22. Ultratech Cement, "Annual report 2021", 2021, https://www.ultratechcement. com/content/dam/ultratechcementwebsite/pdf/financials/annual-reports/ AnnualReport2020-21.pdf.
23. The Center for International Forestry Research (CIFOR) and World Agroforestry (ICRAF) is an organization that envisions a more equitable world where trees in all landscapes, from drylands to humid tropics, enhance the environment and well-being of all. CIFOR and ICRAF are CGIAR Research Centers.

Chapter 3
LIFESTYLE
&
CONSUMPTION

by Christine Loh, Suede Kam
and Amarit Charoenphan

ANALYSIS

by Christine Loh

Changing Behavior to Cut Overconsumption

How we understand a problem determines how we solve it.

We do know that excessive consumption contributes to global climate change and environmental degradation. We also know rich societies consume much more than poor ones. The global poor, or about 12% of people in the world, are those who live on USD 2 or less a day. High-income people are those who can spend more than USD 50 per day. In between, we have those who are in middle to upper-middle-income groups. The poor and low-income groups are under-consuming, as they hardly have enough of the basic necessities.

Those who are middle and upper-middle class in rich countries consume more than their counterparts in emerging economies. For example, research showed the poorest 20% of the people in Britain use five times more energy per person than the poorest 84% of the people in India. Another comparison is 40% of Germany's population are in the top 5% of energy consumers worldwide, compared to 2% of the Chinese and 0.02% of Indians.[1]

Overall, the richest 10% of the world's population consumes 20 times the energy of the poorest 10%.[2] It is clear that to decarbonize, it is the rich world that has to do a lot more, and it is rich people who need to temper their consumption and establish new consumption patterns and desired norms.

Wealthy people — defined as worth USD 30 million or more — are the extreme consumers in environmental terms. While America has more uber-rich people than anywhere else in the world (35%), followed by Europe (28%), Asia is catching up fast (27%).[3] In Asia, the ultra-high net worth people are not only from the higher-income economies of Japan,

Hong Kong, and Singapore but also from the developing economies of mainland China, India, Vietnam, Indonesia, and even Bangladesh, with many of them being billionaires.[4]

The high consumers suck up a lot more energy, water, and other natural resources, including food, than the average consumer in the world. They also have a lot more material things, ranging from clothing, accessories, and gadgets. They are the ones who have big living spaces, multiple homes, and households full of "stuff".

Finding solutions for these excessive consumption groups to alter their lifestyles will have no impact on their real needs but it will help to fight climate change and improve environmental sustainability. Sustainable lifestyles could become a mark of a new identity for humans to reverse the "Anthropocene" in which human action benefits rather than exploits Planet Earth.

Why do people overconsume? Part of the answer lies with how we measure our activities. Economies need to "grow", and we have invented ways to measure "growth". Consumption adds to growth, so the more the better. Adam Smith, the 18th-century philosopher and economist, thought the satisfaction of consumers is the ultimate economic goal and that the economy is fundamentally ruled by "consumer sovereignty" — that is, the desire of consumers.

Another aspect of the answer for over-consumption is that once people can satisfy their needs, they will have many more "wants" that can make them feel psychologically more satisfied. Behaving rich or showing up the Jones creates a sense of emotional satisfaction representing success, or for some, the craving for recognition. In other words, consumption defines one's identity and self-worth.

The conventional wisdom to reduce carbon emissions and environmental footprints is through greening one's lifestyle; in other words, being a green consumer. Changing behavior can do a lot — 72% of global greenhouse gas (GHG) emissions come from household lifestyle consumption, including how we live, eat, and move around.[5] We should start with wealthy people and wealthy societies.

PAST

From Brute to Consumer

Seventeenth-century English philosopher, Thomas Hobbes, famously described life in his days as "nasty, brutish, and short". At the time, the average life span in Europe was between 35 and 40 years, and about the same in China and Japan in the 18th century. Apart from the aristocracies, ordinary people had hard lives and few possessions. The idea of "consumption" did not apply to the vast majority of humans.

The Industrial Revolution of the 19th-century was made possible by using coal for steam engines, which enabled the production of energy and goods at a previously unimaginable speed and scale. The Industrial Revolution created new wealth. The American economist and sociologist Thorstein Veblen, in his famous book *The Theory of the Leisure Class* in 1889, notes that the upper class had accumulated wealth, so they did not need to work. Their social status was all-important, and they were conspicuous consumers. Veblen observed then that this group of consumers was extremely wasteful.

The legacy of the Industrial Revolution is the production of materials goods that became the mainstay of 20th-century culture for the early industrialized economies of Europe and America. Mass production wasn't all bad. It brought down prices of products that many could benefit from. Now, we have become accustomed to an overwhelming cornucopia of consumer goods alongside fashion trends and the world of advertising that gets us to buy stuff. The major problem is the over-exploitation of Planet Earth's functioning boundaries, causing pollution and climate change.

The 20th century also focussed on producing more food using chemical fertilizers and pesticides to help feed the world. Supply chains were created to get food from farm to plate but wasted a gigantic amount along the way through careless logistics. Roughly, one-third of the food produced for human consumption gets lost or wasted.[6] In low-income economies, such as those in Asia, most food loss happens due to limited harvesting capacities and poor storage, transportation, and processing. In wealthy economies, food is wasted often at the consumer's end, e.g., thrown out at supermarkets, restaurants, and at home. Moreover, the nature of global food demand strongly impacts GHG emissions too. To meet the needs of feeding the world, food systems need to provide nutrition while reducing environmental impact. Asia needs to reverse and go in different gears.

Trying "Sustainable Development"

The Europeans may be said to be the pacesetter to change gear. This can be seen through the European Union's tighter environmental standards. Green practices, such as product labeling, waste recycling, insulating buildings well, and prioritizing walking and bicycling as modes of mobility in cities, are all part of using public policy to encourage a green lifestyle. Consumers can participate by buying things that have green labels, separating waste for recycling, cycling rather than driving, and living in well-insulated homes that are more energy-efficient. Other European innovations include a carbon emissions trading system for industry and setting rules to divert capital from high to low carbon investments, commonly referred to as "green finance". While emissions trading does not have a direct impact on day-to-day lifestyle, it is part of Europe's overall low carbon leadership.

In Asia, Japan is the most advanced in waste recycling — in fact, better than Europe — and its energy efficiency standards are second to none. Tokyo's emissions trading system for buildings is an undoubted innovation. South Korea and Taiwan are both quite good at waste collection and recycling too, and China has started to do the same in 2019. As these economies all have manufacturing and agricultural industries, their recycling sectors can turn organic and municipal waste into energy, as well as other products. China started its proto-emissions trading system for the power sector in 2021, and Standard Chartered Bank is offering green mortgages in Hong Kong and Singapore.

PRESENT

Ecological Civilization and Moderation

The problem with doing a bit more of these green activities may just be tinkering at the edges and fostering the illusion that the problem could be solved by consuming more wisely without disturbing the culture of consumption tied to economic growth and the psychology of consumption.

Is there something else that could be done to tap into Asian traditions that could help to transform the underlying culture of materialism that drives consumption? In other words, can "moderation" be popularized and turned into the cultural default in Asia?

The Chinese have come up with the concept of "ecological civilization", which has now been embedded into China's national constitution as a development pathway. This has the potential to go further than the better-known concept of "sustainable development", which calls for human activities to consider environmental, social, and economic impacts so as not to compromise the survival of future generations. Ecological civilization could take things to a higher philosophical level: development should be carbon-free and done through environmental protection and biodiversity.

It is noteworthy that China has made this a political and policy principle. It is the country with the largest GHG emissions and a much-degraded environment as a result of rapid industrialization and urbanization over the last four decades. The concept does provide an alternative development perspective to other emerging economies, however. Moreover, Chinese policy is explicitly focused on "moderation" and not copying the American lifestyle on size and the "drive everywhere" culture. Indeed, the United States (US) has the highest road transportation carbon emissions per capita in the world.

People need mobility, of course, but they do not need cars. This has been clear in Tokyo for years. Tokyo has an extensive, high-capacity, low-cost subway system, and owning a car is expensive. Hong Kong too is a winner, as 90% of daily trips are made by public transport. Many Asian cities now have new underground systems that are efficient and affordable, such as in Bangkok, and Manila is building its Mega Manila Subway. China has good subway systems not only in its largest cities but also in second-tier ones. For hilly cities, Hong Kong's electric stairs going uphill is a valuable innovation.

Asia's famous tuk-tuk three-wheeled taxi services are mostly powered by diesel, a dirty fossil fuel that spews carbon and air pollutant emissions. There are now electric versions on the market, albeit still on the expensive side, but that will change. Asian companies are making solar-powered tuk-tuks of various types for passengers and goods delivery. A good example is Alibaba's auto-rickshaw that holds promise.[7]

To be successful, the notion of "moderation" needs to be internalized by people in the Asia-Pacific region through their underlying traditions since not everything can be done through regulation. East Asia continues to be influenced by the Daoist-Zen-Buddhist traditions of minimalism and thrift that could be tapped for culturally based solutions of less material consumption. The Hindu worldview sees nature as sacred; hence, it favors vegetarianism, and Islam abhors wastage of resources. These are all powerful beliefs to build upon as economies in the region grow in the coming years.

Japanese "cool" includes the KonMari Method started by Marie Kondo to declutter and be liberated from having and keeping too much stuff.[8] Japanese cuisine focuses on quality and presentation in bite sizes. Chinese TCM (Traditional Chinese Medicine) teaches people to eat till they are 80% full, as that is enough. Traditional eating is with the seasons and for health. Gluttony and excess are "out"!

Locally grown organic food is "in". Rooftop gardening in high-rise Hong Kong and Singapore has become fashionable, and some property owners are teaming up with chefs to provide top-quality vegetables and herbs for their restaurants. Hong Kong's Rooftop Republic has 70 rooftop farms,[9] while Singapore's Edible Garden City is an 8,000 m^2 urban farm using a former prison. Farmacy, a startup in Hong Kong, is an early "agri prop" tech company that creates decentralized small, smart mobile farms in cities. High-quality herbs are grown in beautiful cabinets that look good in stores and restaurants, acting both as displays and tiny farms for ingredients.[10] Omipork is the latest rave food in Asia to get people to eat lower down the food chain, and chefs are using their skills to make attractive dishes.

Notably, Singapore is out-doing everyone by promoting a government policy that by 2030, a whopping 30% of the city's nutrition will be locally grown, which is part of the Singapore Green Plan 2030. Technology plays a big role too. Some farms, such as Sustenir Agriculture, are using indoor multi-story LED lighting and recirculating aquaculture systems that produce 10 to 15 times more than traditional farms.[11] Vertical fish farming is also taking off, such as the Apollo Aquaculture Group, which has tank systems on several floors of a building that can purify, monitor, and recirculate water within the farm.[12] Temasek, the city-state's sovereign wealth fund, launched a new entity, the Sustainable Foods Platform, in 2021 to promote such initiatives and invest in relevant businesses.

The Chinese government started a Clean Plate 2.0 campaign to reduce food waste in 2020. During the pandemic, new business ideas have flourished in the food sector. In China, tech companies like Pinduoduo are connecting farmers directly with consumers.[13] In Thailand, the startup, Yindii, uses an app to connect surplus food from supermarkets, bakeries, and restaurants directly to consumers at a discounted price.[14] Indeed, throughout Asia, governments acknowledge the problem, and efforts are underway to reduce food wastage. Greater mobilization and measuring outcomes are the key to show how it can be done.

FUTURE

What's Next?

For a deeper behavior change beyond merely promoting green consumerism, we may focus on the four areas where consumers have a direct impact: food, mobility, fashion, and home.

Food is a good place to develop the alternative vision of the Anthropocene. Food can change mindsets and lifestyles, both from a sustainable and cultural perspective. An Asian cultural approach to what one chooses to eat, how much to eat, and how to prepare food has the capacity to alter our sensibilities that "more" is not "better" just because we have the means to consume. The goal is not consumption but well-being for you and Planet Earth. Moderation is our default as excessive is "crass".

Collaboration between policymakers, businesses, and society in well-designed campaigns is needed. For example, Hong Kong, known as a food paradise, can transform its F&B and hospitality sector by offering "good planet" eating that is healthy for humans and the planet, where food waste is minimal, and portions sizes are designed to suit different appetites, with price options for all pockets. The entire food experience can be digital for those who want to track every atom from farm to stomach. Endangered species are off the menu, of course. Other international eating hubs, such as Singapore and Bangkok, have enormous potential to combine excellent local cuisines with new "good planet" eating too. India, the mecca for vegetarians has a lot to teach everyone.

As for mobility, there is no doubt that subway systems are good for dense urban cities, and Asian cities are building them. Cities can capture a wider range of benefits through good planning that integrates new technologies such as e-tuk-tuks, e-mopeds, and e-scooters. For example, Seoul will replace 100% of its 35,000 motorcycles used for delivery services with e-versions, but it first has to install enough electric charging facilities. In Chinese cities, internal combustion engine motorbikes were banned years ago, having been replaced by electric ones.

Even with e-mobility, one does not want so many vehicles that they end up clogging the roads. This can only happen by skillful planning coupled with digitalized traffic management to regulate the use of road space — a public resource — and, yes, road pricing is a feature of the future.

City authorities need to make walking and biking attractive and safe, taking into account the challenge of equatorial, tropical, and sub-tropical cities, where temperatures are high in many parts of Asia. Singapore, a flat city, is creating Cycling Path Networks to enable inter-town mobility, and Hong Kong, with its hilly terrain, has long cycling paths in the New Territories along the coast. People will get fitter, if walking and cycling become possible in more and more cities — another socio-economic benefit.

The challenge is longer-distance travel. The next generation of high-speed trains using maglev technology makes trains competitive against aeroplanes on the aspect of time. For example, the fast train between Beijing and Shanghai takes about 4.3 hours, but a maglev train can cut the time by half, which is what a flight takes. When China increases its non-fossil fuel usage to power transportation, fast trains on heavily used routes beats driving or flying in terms of carbon emissions. Moreover, cross-border travel within Asia is on the cards. A new fast passenger train between China and Laos opened in 2021, a project that could eventually connect a good part of Southeast Asia. Public policy will be needed to use price to encourage the most environmental-friendly means, which may well be via rail rather than flying.

So what of our desire to see the world? The meta-universe helps us to see Everest without physically going, and it can be an enhanced educational experience. Technology allows us to experience amazing things and learn about them at a sufficiently deep level. We will understand that moderation means that not all of us can visit places like Dunhuang and Angkor Wat, but technology will help us learn about them. Yes, visits to even our local natural marvels may need to be restricted or rationed for the betterment of all. Cutting tourism, in the traditional sense, is a way to decarbonize, along with fewer business trips since we are now used to e-meetings. This will have an impact on hospitality related jobs as we see those jobs today, however. There is no easy answer. Decarbonization efforts cannot happen without deeper, structural changes to our societies. Again, consumer moderation can set the pace for change over the next two decades as we move closer to 2050 and feel greater pressure to decarbonize.

In an age of self-expression, the fashion industry is stuck with brands selling the same things everywhere. Shockingly, it is the second-most polluting industry in the world after the oil industry, as making textile and clothing require very large amounts of many resources. While there are arguments about how one measures the carbon footprint of such a complex industry, researchers have shown that fashion contributes 2% to 10% of the world's carbon emissions. Fast fashion — producing inexpensive clothing rapidly in response to trends — is "out". It is extremely wasteful of resources because "fast clothes" are quickly discarded as well. It still makes sense for mass retailers because resources and labor are priced too cheaply. Fast fashion has become the poster child for extreme wastefulness.

Asia's pioneering Hong Kong Research Institute of Textile and Apparel[15] has some answers to that. It has invented new materials made from coffee bean waste, as well as ones that can absorb carbon. Its technology to recycle textiles, already taken up by some of the international brands, offers us a glimpse of the future. Recycling is only one aspect of the challenge — moderation requires us to consume less but use what we have more intensely. Is there a business model where clothes can be continuously renewed, updated, and transformed? One way or another, the industry has to find a solution. The commitment of more than 130 retailers, with the likes of Burberry, H&M, Nike, or Inditex, at COP26 to halve their 2019 emissions by the end of our decade is a sign that there will be at least an effort towards that. The problem to solve lies with new business models that are not focussed on greater and greater consumption.

The construction of new buildings is highly carbon intensive. Retrofitting buildings is less so. Existing buildings can be refurbished to fit new circumstances for living and work. COVID-19 has already driven new investment thinking in properties. There is no shortage of new technologies and management methods to make buildings more energy efficient, which is especially relevant for expensive properties since their occupants consume much more than those in low-income housing. Typical Asian urban housing is generally high-rise and smaller than their counterparts in Europe and America. The future lies in the better design of space use and common facilities. Singapore's public housing provides a guide for the world on how high-density living could be pleasant, convenient, and investible.

India, too, has good ideas and practices for sustainable buildings that take its climate, traditional design, and science into account. Perhaps few people outside India know that the Sohrabji Godrej Green Business Centre, built in 2004 for the Confederation of Indian Industry, is the icon of India's green building movement. Not only was it the first LEED Platinum Certified Building in India and outside the US, it is also a net-zero building. The architect, Karan Grover, may be said to be India's father of green architecture, and there are now many green architects and green buildings throughout the country.

Further afield is the Middle East — the region synonymous with oil and gas. However, the petrostates are preparing for a post-fossil fuel era by investing in sustainable living and new energy technologies. A symbol of this effort is NEOM, a newly planned, green, smart city in Saudi Arabia powered entirely by renewable energy, the first phase of which could be finished by 2025. The name "NEOM" combines "neo", which in Greek means new, and "M" represents "mostaqbal", which means future. We shall witness over the coming years whether this project can fulfill its mission to create a desirable example of not only decarbonized urban development but also sustainable lifestyle and consumption.

CONCLUSION

Call to Action

Getting to net zero needs a revolution, not only to consume a bit "greener" but to change our whole conception of consumption. Those who see themselves as trendsetters and influencers can give themselves a special mission to make "moderation" appealing, one that can satisfy at a deeper level. In other words, we need to make overconsumption "uncool". In a revolution to change how the world functions, the role of governments is critical. They need to work with businesses and communities to make low-carbon choices work, such as by walking more, restricting certain foods practices, stopping transporting fancy bottled water around the world by making them obscenely expensive, giving policy priority to public transport over cars, and changing building codes to massively favor green designs. Only public policy can incentivize green outcomes and penalize unsustainable ones, so governments must act. For zero and low-carbon choices to work, entrepreneurs who are dreamers have the role to envision new business models. Those in communication can help to promote the needs for the revolution. There is a lot of excitement from people who see the future — it is time to collectively give it a go in our societies.

STARTUP CASE STUDY
OmniFoods

by Suede Kam

The plant-based food industry is booming. Consumers worldwide now have a variety of alternative protein choices. The strong growth is linked to a dietary shift away from excessive consumption of traditional meat products because of environmental and health concerns.[16]

In Asia, OmniFoods is converting pork lovers into flexitarians who are open to reducing meat from their diets. Headquartered in Hong Kong, the food tech startup created the world's first plant-based pork and branded it OmniPork. The meatless pork is made of shiitake mushroom, non-GMO (genetically modified organism) soy, pea, and rice. It is dubbed "cruelty-free, vegan, Buddhist-friendly, and non-GMO [with] no cholesterol, added hormones or antibiotics".

Infused with east and southeast Asian cultures, pork has long been a culinary preference in Asia-Pacific. The region accounts for more than half of the pork consumption in the world.[17] OmniPork is a natural business pursuit to provide a pork alternative that also addresses the climate crisis. After all, livestock production emitted 14.5% of global GHGs, according to the United Nations Food and Agriculture Organization, while others claimed it could be as high as 51%.

Pig meat sees the second-highest total emissions after beef meat. It is responsible for 11% of animal agriculture emissions amounting to 0.84 gigatons of CO2 equivalent.[18] Consumers can help curb up to 61% of that emission by opting for plant-based diets.[19] With a vegan pork analog, OmniFoods can chip away 1 to 3.4% global GHG emissions at its maximum impact.

OmniFoods focuses on growing the population of flexitarians, sometimes called "meat reducers" or "semi vegetarians". One will find it less likely to relapse from such a flexible lifestyle as compared to a stricter diet like veganism — the catch to sustaining any new diet is convenience.

OmniPork comes in different forms, minced, strip, patty, and luncheon meat, as seen at over 40,000 locations in 20 countries. The product line further expands into instant meals, which are available not only in grocery stores but also in 7-Elevens in Hong Kong and FamilyMarts in Taiwan. These ready-made meals come in local Asian flavors and are available around the clock at the abovementioned convenience store networks.

Convenience food is generally perceived as unhealthy. However, OmniFoods has selected the most popular processed meat in Asia-Pacific and taken on this perception head-on — enter the OmniPork Luncheon, the world's first vegan luncheon meat. As compared to traditional luncheon meat, the OmniPork Luncheon contains 49% less fat, 62% less sodium, zero cholesterol, zero MSG, and no carcinogenic nitrates as preservatives.

In Hong Kong and Macau, McDonald's has launched a six-item menu featuring an exclusive thicker cut of OmniPork Luncheon. It is by far the biggest collaboration between a plant-based meat company and quick-service restaurants in Asia-Pacific.

OmniFoods talks about versatility a lot. Its Canada-based R&D team develops properties based on Asian eating cultures and cooking habits. In addition to McDonald's menu, OmniPork can be seasoned and prepared the way regular pork is in ethnic cuisines, such as in tacos, gyozas, sliders, and spring rolls.

This way, it is versatile enough for chefs worldwide, from fast food to Michelin-starred restaurants, to alchemize new menus. Putting OmniFoods products in more touchpoints will increase the frequency consumers encounter their products and brand mission. In turn, the exposure will help drive awareness and propel meaningful lifestyle change.

Three elements placed this Asian startup on a hypergrowth path.

(1) OmniPork found exactly where its growth opportunity was. It blazed a new trail when other brands like Impossible Meat and Beyond Meat focused on beef. In 2021, the Omni Seafood line was born, offering the world's first non-breaded plant-based fillet.

Asia-Pacific is home to two-thirds of the global seafood consumption, and the non-breaded fillet is more accepted by the Asian palate. According to OmniFoods' CEO David Yeung, "it [was] a matter of proposition and filling the gaps". The market foresight that is characteristically regional and data-driven has reinforced OminFoods' leadership position in Asia-Pacific.

(2) OmniPork identified partnerships that could scale fast with the least resources possible. For instance, McDonald's luncheon meat menu resulted in mass advertising for OmniPork at over 230 outlets overnight. OmniFoods often leverages first-mover advantage to dominate retail and distribution networks, where accessibility and mindshare rise exponentially because their products are carried in major chains such as Woolworths, Sainsbury's, Starbucks, McDonald's, Taco Bell, Four Season Hotels, Alibaba's Tmall, and more.

(3) OmniPork made consumers' purchasing decision simpler. OmniPork has achieved price parity with pork in Hong Kong. Without a price deterrent, consumers have one less variable to consider and a greater incentive to select pork alternatives and lifestyle changes. It is no longer a false option limited to the wealthy and those who can afford it as a trade-off. If a consumer's primary concern is rooted in animal welfare and environmental impact, then the choice should now be almost binary.

The plant-based meat industry has heated up because venture capitalists have grown wary of regulatory issues of cell-based meat, a direct competitor. OmniFoods' mother company, Green Monday Holdings, has bagged USD 70 million in 2020, led by TPG and Swire Pacific, and is seeking USD 100 million in funding.

Currently, all of OmniFoods' production takes place in Thailand. A second plant is being built in Guangdong, China, though it is actively surveying a third site in Taiwan — it looks as though supply cannot meet demand. Hence, OmniFoods' next step is to optimize production efficiency and perfect the unit economics in various markets. Short of full-fledged production, OmniFoods is already challenging the economics of real meat rivals and has the potential to price its products even more competitively.

The profit margin in this industry still has room to grow and will thus attract competition; however, newcomers will need to overcome the competitive advantage in established distribution networks and intellectual properties that Omnifoods, which has been around since 2012, currently enjoys. Such potential competition will encourage innovation and diversity in offerings and price points and allow the industry to flourish.

Asia-Pacific's appetite for meat will continue to rise, strongly influenced by growing affluence in countries like Vietnam, Indonesia, Japan, and China. The adoption of plant-based meat depends on how brands associate with this group in terms of status and moral binding. Another force to be reckoned with is youth. As the world's youngest region, Asia-Pacific has the power to shape and dictate future consumption patterns and lifestyles. Motivations and barriers differ; more regional players who are well versed in cultural sensitivity will pop up to accommodate the demand in different markets.

To accelerate flexitarian lifestyle adoption, some have argued that it is necessary to improve the alternatives to the point they are indistinguishable from real meat. However, when faced with the next African swine fever or pandemic that results in culled livestock, would taste be a priority? When grappling with a growing population and insufficient land to meet our food needs, would taste be a necessity?

Plant-based technology promises to use less water, land, and energy to produce food sources. It lowers GHG emissions. Of course, it mitigates food safety and food security crises too. Ultimately, we consumers hold the key to the problem. Whether or not OmniFoods or a similar startup offers comparable alternatives to unsustainable animal products, it is up to us to buy in and change our habits today. The next time we purchase our meat, the choices we make will help decide our future.

CORPORATE CASE STUDY
L'Oréal

by Suede Kam

The L'Oréal Group is the world's largest beauty manufacturer, ahead of Unilever and Estée Lauder.[20] With 35 brands, L'Oréal reaches 1.2 billion customers across 150 countries each year. In Asia-Pacific, L'Oréal's sales top the industry and leave its closest regional competitors, namely, Shiseido, SK-II, and Amorepacific, in its wake.

It is no wonder L'Oréal's carbon footprint is significant in Asia-Pacific. A third of L'Oréal's sales and one-tenths of its production happens within the region. Asia's Scope 1, 2, and 3 emissions are estimated at 3,563.8 thousand tons.[21] Despite it being a French company, its carbon footprint across Asia-Pacific, like its sales, surpasses its regional contenders.

The broad-based footprint implies opportunities for improvement. In fact, L'Oréal has shaved 81% off its carbon emissions since 2005. In the past few years, it has been minimizing its carbon footprint by 25–35% year-over-year without compromising production capabilities. The production volume has gone up each year due to growth in Asia and emerging markets. Even with heavy decarbonization initiatives and investment, L'Oréal's operating profit has maintained a high teen percentage (between 18 and 19% in the last 5 years). The consistent performance can be explained by innovation and market-led growth as well as rigorous cost control.

L'Oréal's target has been to reduce 100% of its carbon emission from its 2005 level. After 17 years on this journey, they are not far away. How can Asia-Pacific help with the last mile?

In 2019, China became L'Oréal's first market to realize carbon neutrality. Both its plants in Yichang (the largest makeup production center in Asia-Pacific) and Suzhou and its distribution centers, research and innovation center, and offices are fully carbon neutral through improving energy efficiency and tapping into renewable energy sources.[22] A year later, another plant in Baddi, India, qualified as well. Asia-Pacific has demonstrated how a low carbon future can look like and serves as an example for global operation sites to follow.

Scope 3 is where the heavy liftings are done — it represents 99.7% of L'Oréal's total emissions. Over half of it traces back to beauty product formula sources. The choice of ingredients determines how raw materials are extracted and prepped, how products are designed and manufactured, and how consumers use the end products. Much is done behind the scenes; however, consumers will be the ones to see, smell, and touch them in the end.

For transparency, L'Oréal reveals product environmental and social profiles on its own "Product Impact Labeling" system. In late 2022, such information will also be found on "Eco Beauty Score", a similar rating system designed for industry-wide usage. A total of 8 out of the world's 10 largest beauty companies, namely, L'Oréal Group, Unilever, The Estée Lauder Companies, P&G, Shiseido, Natura & Co, Beiersdorf, and LVMH are joining forces to co-develop the system to assess the environmental impact of beauty products, from formula to packaging, to usage.

That's groundbreaking. Currently, there is no standard way to compare products across brands. A total of 36 companies will form a consortium in this effort to make sure consumers are well informed and enabled in the buying process. Leading Asian brands Shiseido, Amorepacific, and KAO are on the roster, too.

This is a powerful tool to capture carbon reduction. Consumers can indicate what raw materials they disapprove of, and through their purchasing behaviors, influence how compositions and products are made. Having contributed to a third of L'Oréal's sales, Asian consumer preferences will play an important role to displace carbon-intensive materials and products.

Taking the second place is packaging — 12% of the total emissions. While 60% of all packaging is plastic, L'Oréal plans for it to be refillable, reusable, recyclable, or compostable by 2025. The next carbon-intensive area pertains to the end-of-life treatment of sold products. Consumers are partially responsible for this 5.4% of L'Oréal's emissions. The decarbonization opportunity is tied to how packaging or product waste is disposed or recycled. In Asia-Pacific, L'Oréal has launched cross-brand recycling programs to collect used product containers from 13 brands. Major barriers to recycling are specific to individual markets, so L'Oréal has collaborated with various local social enterprises to navigate each scenario.

The fourth major carbon contributor is transportation and distribution associated with finished products to the first customer's delivery point and home. It produces 4.6% of L'Oréal's emissions. In Tokyo, L'Oréal introduced cargo bikes for delivery. It guarantees zero emissions but is not the most efficient. Over in Shanghai, L'Oréal gradually lowers the carbon emissions by adopting electric, biofuel, or liquefied natural gas (LNG) vehicles. This constitutes 70% of downtown deliveries in partnership with the largest Chinese courier network. In Singapore, electric vehicle delivery is available across 40 stores in hopes of cutting 50% of carbon emissions each month.[23] Like in China, the Singapore team works with a local transportation partner to ramp up the delivery capacity. With successful pilots, L'Oréal plans to expand the sustainable transport initiative to 50 cities by 2025.

E-commerce today has experienced unprecedented growth — the COVID pandemic has reinforced habits of online purchasing.[24] The trend will continue and raise the number of last-mile deliveries. There will be a case to accommodate these deliveries, which has great decarbonization potential as mobility technologies are becoming more affordable, reliable, and carbonless.

Beauty product consumption in Asia-Pacific is experiencing strong growth, driven by Chinese consumer demand for luxury products. Across various L'Oreal's initiatives, Asia-Pacific consumers now have the avenues to contribute to decarbonization.

Initiatives are implemented not only through its premium brands which tend to have higher margin, stronger brand affinity, and more environmental-conscious customers, but also brands that cover a large consumer base. L'Oréal takes a systematic approach and rolls out its decarbonization initiatives city by city. It effectively gathers feedback and helps manage risks. Consumers hold a renewed impetus for sustainability and call for change towards the environment. Consumer preferences and behaviors are shifting, so beauty brands need to listen closely.

By 2025, L'Oréal will achieve carbon neutrality in all operating sites. By 2030, L'Oréal will halve its carbon emissions and turn fully carbon neutral. By 2050, L'Oréal will be net zero. This competitive advantage will favor those who can remain competitive, resilient, and sustainable — all at the same time.

INTERVIEW
Soranun "Earth" Choochat
Chief Executive Officer, ETRAN

by Amarit Charoenphan

Across the world, one of the biggest concerns from a sustainable lifestyle perspective is the amount of trash and carbon footprint that we have accumulated from all our e-commerce and food deliveries. To find out more, I had a chance to interview Soranun "Earth" Choochat, CEO of ETRAN (https://www.etrangroup.com) — Thailand's leading electric motorcycle company that is expanding its business model from direct to consumer to B2B and from selling bikes to renting them. He is doing the unthinkable, building a homegrown electric vehicle (EV) brand and infrastructure in an emerging market from nothing, and he has now taken his company to Series A and 5,000 orders after working on the business for 5 years. His model also aims to incorporate carbon footprint tracking and carbon credit trading solutions directly on the motorcycle, and I caught up with him to find out more about how he was doing.

Earth, can you tell me more ETRAN?

At ETRAN, our goal is to be the cleanest mobility company, with more than 5,000 unit reserved (as of May 2022). To deliver on our promise, we design, manufacture, and deliver to our customers high performance electric motorcycles using two business models — sale and rent. Now the growth from rental is growing very fast. We see a demand of more than 5,000 units by year-end. We are also not just selling the electric motorcycle, we are investing in scaling our powerstation ("the swapping station") across Bangkok to create high accessibility for our customers to swap anytime they need, as well as providing after-sales service, an IoT platform for fleet management, and battery service too.

From the darkest experiences of almost going under, to now becoming a leading player in the industry, what was the pivotal moment for ETRAN?

To grow from a startup into a technology business, raising funding and scaling was the pivotal moment for our company. We are a bootstrapped company with the support from many angel investors, but at a certain point we need to raise significant funding to get production and the ecosystem. We don't specifically look at venture capitalists, but being strategic investors in the hardware business requires a lot of patient capital. Venture capitalists that I pitch to only ask for reports, while the corporates that invest in us are more like a cofounder who gives value to the company in terms of know-how and experience, partnerships, distribution, and other synergies such as ERP, accounting, and a back office. By bringing their full team to work together, our corporate investors have helped us mature into a full-fledged business much faster than had we did it on our own.

What are the problems that you are trying to tackle and how has that evolved as your business grew?

As our dream has got bigger, we want to be one of the world's first Environmental, Social, and Governance (ESG) electric motorcycle companies. Our vision now in 2022 is to be the ESG leader in Thailand's electric motorcycle industry by achieving 10,000 units (25% of the market share) plus 20,000 tons of carbon reduction and empowering our 10,000 riders to reduce their carbon footprint by riding our motorcycles. To achieve our environmental goals, we need to improve manufacturing processes, including how we manage waste and the carbon footprint of our supply chain. We plan to utilize new manufacturing and assessment technologies to track and reduce our carbon footprint. On the social front, we want to empower our riders to be climate warriors by allowing their rides to generate carbon credits that we will distribute to them via blockchain. On governance, as we are ramping operations and building a new factory next year, we need to move toward using ERP to help us manage our business better.

As the EV industry matures, what was the market's response to your electric motorcycles and how did you pivot or persevere?

Initially, we started with selling high-performance motorcycles, which were quite expensive due to the low production volumes. However, during the pandemic, we noticed that more and more rides were using their private vehicles to make money as delivery men and women, which was concerning, as more bikes on the road mean more pollution. That was when we realized that our bikes could expand to this emerging market of an end-to-end fleet solution, in that we could rent our bikes to riders so that they would not have to spend a lot of money to buy their own. We would also build an ecosystem of swapping stations, IoT, and management services to help make it easier for consumers and fleet operators to manage their vehicles using our proprietary vehicle management platform. As the demand has surged, we are continuing to A/B test various solutions such as white labeling our entire ecosystem or being an OEM bike manufacturer to see how we can continue to grow with our corporate partners.

Who are your largest market and target customers?

While ETRAN has shifted more from B2C to B2B, demand is still strong in both. We operate in a direct-to-consumer model by selling online with no dealers like Tesla. B2B and rental subscriptions models are an exciting prospect for us as it allows us to have long-term recurring revenue while staying ahead of competitors, as consumers will be able to afford our EV motorcycles easier than more expensive competitors. As we build scale and credibility, we are exploring B2G models where municipalities and provincial and national governments are looking to roll out EV fleets for motorcycle taxis, which are prevalent in Asia.

How do you quantify ETRAN fleet's carbon reduction?

We're in the process of exploring how to generate carbon credit from our fleet, because for every 1 km ridden by ETRAN, we already know that we are reducing 45 g of carbon emissions. Clearly, if we can quantify and generate renewable energy credits, especially if our battery swapping stations are also powered by renewable energy, we believe that we can trade these carbon credits not just in Thailand but other countries as well. We are also exploring ways to quantify our other environmental footprint; for example, for Tesla and Ford, the manufacturing of their vehicles use a lot of water. If we can track the environmental footprint for our entire production, we will be able to become truly the cleanest mobility company in the world.

What are your operating business models and where are you planning to go next?

ETRAN is a company owned and operated in Thailand, but we have now started to expand into new vertices with ETRAN Rental, our subsidiary in charge of vehicle rentals that is already operating — 90% of our bookings come from this model. Furthermore, we are expanding internationally with ETRAN Malaysia as our first overseas destination due to our strong investor base. While all our motorcycles are sold in Thailand, and the majority of the business model is B2B, we believe we will be selling more bikes to the general consumer over time. Furthermore, we are also branching out into swapping stations where we currently have four stations to swap batteries and provide maintenance on our electric bikes. We are also now partnering with gas stations and other interested parties to become our swapping station partners and hope to deploy 16 battery swapping stations by the end of the year.

After Thailand, what are your Asian ambitions?

Our goal is to be the largest electric mobility company in Southeast Asia in 5 years by selling over 1,000,000 vehicles, which we are highly confident of achieving. In order to get there, the major roadblock that we need to overcome is consumer adoption of electric mobility vehicles, which for every type of vehicle, whether it be a commercial vehicle to a lifestyle and recreational vehicle, the number one concern is still the infrastructure across the region. In Thailand alone, we estimate that we will need to build 10,000 battery swapping stations at the tune of 1 million baht per location to service 120 vehicles a day per station. The task seems monumental, but we believe that with our subsequent funding rounds, we will be able to raise over USD 300 million for this infrastructure rollout to become a Thai electric mobility "unicorn".

What are some of your Top Highlight projects?

We currently are one of the strategic partners providing electric motorcycle fleets to Robinhood, a food-delivery app that is backed and ventured-built by Siam Commercial Bank, Thailand's oldest bank which has become one of the top food delivery apps in a short period of time. Currently, beyond food delivery, we are talking to many logistic partners to power their fleets.

What are your biggest challenges ahead?

Building EVs and anything hardware-related is difficult, and doing that in a country where no one has done it before is an extreme challenge that we are ready to take on. Currently, one of the biggest challenges to becoming the cleanest mobility company in Asia is hamstrung by the fact that sustainability in the supply chain is low, especially during COVID-19, where everything took longer as we had to source for components during this time. The chip shortage not only affected us but even large companies like Tesla and Apple. Startups live and die on speed, and we know that if we are not moving quickly, regional and global players are ready to enter the Southeast Asian market. However, more competition is a sign that we are in the right sector, since with a big enough market, there will naturally be competition. The big players are proof that there is a market, like Clubhouse being cloned by Twitter and Facebook.

Other challenges are quite similar with any deep tech ventures: we need to set up an R&D lab to have our own facilities so that we can develop and control more inclusive tech on our own, like the carbon credit system and battery swapping technologies. This is where government policy and support is so important, because we will need to make a lot of investment to build the supporting infrastructure to support EV adoption like charging facilities, lowered taxes on components, machinery, the purchasing of EVs, and so on.

On this front, the Thai government is headed in the right direction but at the wrong pace. In order to maintain our leadership as the top 10 vehicle manufacturers in the world, we need to be more aggressive to promote this world-changing industry. Furthermore, the government will also need to solve one of the biggest bottlenecks in the Thai automotive industry, which is human capital, as there are very few people who have the experience and know-how to create EV vehicles, let alone assemble one. There is a massive talent gap, and the effort it takes to develop one is very challenging, one that ETRAN cannot do alone.

Can you tell me about your fundraising and what are your plans?

Currently, we are valued at USD 10 million in our Series A this May, and we are currently raising our Series B at a USD 30 million pre-money valuation with major developmental institutions and sustainable funds.

With the story of ETRAN, we are seeing a dynamic across the region to make sustainable transportation, not just for personal use, but for the consumer's expectation that the businesses they buy from will care for the environment and deploy a fleet of EVs. Whether it is electric delivery fleets like ETRAN, motorcycle taxis by the Electricity Generating Authority of Thailand (EGAT), or electric tuk-tuks like MuvMi, the B2B and B2G space is the stepping stone to mass adoption by consumers, and the trend will only accelerate from here. Particularly in Southeast Asia, where many countries are pledging to ban ICE vehicles by 2035–2050, expand charging stations, and ramp up EV production domestically, I expect that consumer adoption of a vehicle-less or vehicle-light lifestyle by relying on on-demand rentals, car sharing, or leasing of EVs will only accelerate as countries position to become a leader in this massive automotive technological shift.

In summary, how does ETRAN's solutions create a significant impact in contribution to a more sustainable lifestyle and consumption?

At ETRAN, we are driven to create a better, cleaner, if not the most efficient and most equitable electric motorcycle. Not only do we produce no carbon dioxide, but we are also always looking at ways to green our entire vehicle and supply chain, starting with the use of recycled plastics in our body parts. We aim to create a heavy-duty, yet high-performance, commercial delivery fleet that has carefully engineered batteries and controllers optimized for a high-temperature, congested environment starting from Bangkok. But where I see ETRAN making the most impact in terms of consumer consciousness is our business model, which is to not sell the bikes, but to lease them. If we sold all our bikes, we believe that due to the current cost structure, only rich consumers would be able to afford them but not be likely to use them daily, which defeats the purpose of climate change mitigation where we are trying to replace traditional motorcycles. We want to flip it so that real users, who ride them daily for hundreds of kilometers a day, can enjoy cost savings, cleaner air, and an opportunity to do good every day. Once we are able to bridge the digital and technological divide gap in Thailand's society, we believe that these power users will spread the word about the durability performance and attractiveness of EV bikes, and will be proud to tell all their clients and friends that they are saving the world on every food delivery, e-commerce purchase, or motorcycle taxi ride. These are the most demanding consumers that we are building our fleets for, and if we can win them over, we will not just change how Thais ride to work, but we will have millions of brand ambassadors across the Southeast Asian region that now understand not just what EVs are, but the meaning of carbon footprint reduction.

Figure 3.1.   ETRAN's electric delivery fleet of motorcycles.

Endnotes

1. University of Leeds, "Shining a light on international energy inequality", Faculty of Environment, University of Leeds, 16 March 2020, https://environment.leeds.ac.uk/faculty/news/article/5311/shining-a-light-on-international-energy-inequality.
2. Ibid.
3. CNBC, "Crazy rich Asia: Home to the fastest-growing pool of wealth in the world", 6 September 2018, https://www.cnbc.com/2018/09/06/crazy-rich-asians-wealth-in-asia-grows-faster-than-us-europe-in-2018.html#:~:text=However%2C%20Asia%20is%20quickly%20closing,net%20worth%20countries%20in%202017.
4. CNBC, "Asia-Pacific has more billionaires than any other region, as pandemic boost wealth", 7 October 2020, https://www.cnbc.com/2020/10/08/asia-pacific-is-home-to-most-billionaires-globally-pandemic-grows-wealth.html.
5. Kimberly A. Nicholas, "Changing behavior to help meet long-term climate targets", World Resources Institute, https://www.wri.org/climate/expert-perspective/changing-behavior-help-meet-long-term-climate-targets.
6. UNEP, "Worldwide food waste", https://www.unep.org/thinkeatsave/get-informed/worldwide-food-waste.
7. Michal Toll, "Awesomely weird Alibaba electric vehicle of the week: $1,691 solar-powered tuk-tuk trike", electrek, 8 May 2021, https://electrek.co/2021/05/08/awesomely-weird-alibaba-electric-vehicle-of-the-week-1691-solar-powered-tuk-tuk-trike/.
8. The KonMarie Method, "What is the KonMari Method™?", https://konmari.com/about-the-konmari-method/#:~:text=The%20KonMari%20Method%E2%84%A2%20encourages,that%20no%20longer%20spark%20joy.
9. Sustenir, https://sustenir.com/.
10. Farmacy HK, https://www.farmacyhk.com/.
11. Singapore Food Agency, "Strengthening our food security", https://www.ourfoodfuture.gov.sg/30by30.
12. Justin Tan, "The future of agritech: Inside Singapore's vision for food security", GovInsider, 9 July 2021, https://govinsider.asia/smart-gov/the-future-of-agritech-inside-singapores-vision-for-food-security-melvin-chow/.
13. Pinduoduo, https://en.pinduoduo.com/stories#agriculture.
14. Yindii, "Welcome to Yindii", https://www.yindii.co/.

15. Hong Kong Research Institute for Textile and Apparel, https://www.hkrita.com/.
16. NielsenIQ, "The F word: Flexitarian is not a curse to the meat industry", 2019, https://nielseniq.com/global/en/insights/analysis/2019/the-f-word-flexitarian-is-not-a-curse-to-the-meat-industry/.
17. Organisation for Economic Co-operation and Development, "Meat consumption (indicator)", OECD Data, https://data.oecd.org/agroutput/meat-consumption.htm. Accessed: 26 March 2022.
18. Food and Agriculture Organization, "GLEAM 2.0 — Assessment of greenhouse gas emissions and mitigation potential", https://www.fao.org/gleam/results/en/. Accessed on: 26 March 2022.
19. Zhongxiao Sun, Laura Scherer, Arnold Tukker, *et al.*, "Dietary change in high-income nations alone can lead to substantial double climate dividend", *Nature Food* **3** (2022) 29–37, https://doi.org/10.1038/s43016-021-00431-5.
20. WWD, "Beauty's top 100", May 2020, https://wwd.com/beauty-industry-news/beauty-features/top-100-beauty-brands-2020-beauty-inc-1234805760/.
21. Universal Registration Document 2021, https://www.loreal-finance.com/eng/universal-registration-document.
22. L'Oreal, "China becomes L'Oréal's first market realizing carbon neutral with operation sites covering plants, distribution centers, research & innovation center and offices", 9 September 2021, https://www.loreal.com/en/china/articles/commitments/china-becomes-loreals-first-market-realizing-carbon-neutral/.
23. Amanda Lim, "Cleaner footprints: L'Oréal aims for greener deliveries in Singapore with latest sustainability push", *CosmeticDesign-Asia*, 4 May 2021, https://www.cosmeticsdesign-asia.com/Article/2021/05/04/L-Oreal-aims-for-greener-deliveries-in-Singapore-with-latest-sustainability-push.
24. United Nations, "Global e-commerce jumps to $26.7 trillion, fuelled by COVID-19", UN News, 3 May 2021, https://news.un.org/en/story/2021/05/1091182.

Section 2
CITIES

INTRODUCTION

by Roman Y. Shemakov

Cities are the most fundamental organizing unit of the modern world. They bring people together, direct global finance, concentrate knowledge, breed change, and support every function of communal survival. Simultaneously, they overconsume most of the world's energy, building material, and natural resources. In this regard, cities have the most potential to address global emissions. Whilst cities take up less than 3% of the world's land area, they use up almost 78% of its energy, and account for more than 60% of global emissions.[1]

As extreme weather events continue to proliferate, it will become increasingly more difficult to stabilize the ecosystem. In 1950, 30% of the world's population lived in cities. Now, it is 55%. In the next 30 years, that number will grow to 68%.[2] Beyond the fact that the majority of modern infrastructure is unprepared for this large influx of people, our climate is even more unprepared for cities running at their absolute limits. There is only so much the dam can hold, in all three senses of the word. If we continue to rely on analog energy, building, and transportation technologies, CO2 output will remain exponentially tied to population growth. More people in cities mean more strain on existing infrastructure, worsening metabolic control, and a lack of consideration for externalities. In the most optimistic sense, course-altering changes can be made through each of these nodes.

Asia-Pacific is the region with the greatest opportunity to break this cycle. The continent is home to most of the world's urban population but has almost the lowest level of urbanization.[3] This means that by 2050, almost 70% of the world's urban growth will happen in Asia.[4] Most significantly, the fastest-growing urban centers have fewer than 1 million people. While one in eight people live in 33 megacities worldwide, close to half of the world's urban dwellers reside in much smaller settlements with fewer than 500,000 inhabitants.[5]

This creates incredible opportunities for addressing climate change at its source. Most of the largest cities in the world are imprisoned by their infrastructure. Uprooting and transitioning the energy, building, and transportation systems of contemporary global cities is a worthwhile task, but it will be an uphill battle. On the other hand, the infrastructure of the fastest-growing cities is in its relative infancy. These cities can leapfrog the degraded machinery of their older, behemoth counterparts. Constructing carbonless energy, buildings, and transportation systems is possible.

Transforming our Energy, Built Environments, and Transportation

The great news is that a significant amount of technology needed to make our urban areas carbonless already exists or is close to becoming viable. The following three chapters delve deeply into the existing infrastructure problems, how to fix them, and where resources and energy ought to be directed.

In "Energy", Sandro Desideri and Alberto Balbo outline how to move out of humanity's "code red". They soberly describe Asia's need to technologically leapfrog the evolution of energy systems witnessed in the West and dramatically accelerate the adoption of sustainable energy sources. At the same time, it is vital to acknowledge the reality that several Asian countries are still emerging economies that rely on polluting or subsidized energy. In this quest, Asia needs to combine distributed grids, improved efficiencies, renewable energy, and carbon capture. The production of solar, wind, and green hydrogen is becoming cheaper and more efficient. Incremental innovations in lithium-ion battery storage are being coupled with unorthodox methods of compressed-air storage, gravity storage, and pumped hydro. The chapter provides a fundamental call to action: the availability of technology is not matched by a financial system or a robust set of policies holding entities accountable for their net zero pledges. A significantly more ambitious roadmap is required to be willing to destroy and create industries based on the cost to the planet's 2-million-year metabolic cycle, and not on nascent flash crashes, insider trading, digital boosters, and wild west securitization.

In "Built Environment", Eric Chong analyzes how we must limit our buildings' impact on the surrounding ecosystem. The built environment accounts for 40% of all greenhouse gas emissions. First, the chapter considers the buildings, roads, utilities, and supply chains that connect our cities. Most importantly, the chapter emphasizes the technology that is necessary to reduce waste and transition quickly. The section dissects the emerging urban technologies (5G, IoT, AI) that will be necessary to track urban data flows, understand the cumulative impact of the built environment, and address its problems at the source. Finally, the chapter points towards a world in which legislative frameworks foster jurisdictions where urban experimentation is possible, meaning that long-term (300–500 years) contribution to global knowledge is valued over short-term (50–100 years) rewards.

In "Transportation", James Kruger and Davide A. Nicolini look at the needed transformations to ways we move through our cities. Overall, transport represents 9% of the global GDP and is the primary artery that allows cities to exist. The chapter diagnoses the problem: the damage of emissions to the ozone layer, respiratory health, the healthcare system, and planetary flows. The chapter then analyzes the experimentations in conveyance with electrification, car-sharing, robo-taxis, effective public transportation, and improved batteries. Kruger and Nicolini conclude by looking at aviation, shipping, buses, and trucks and consider both the energy and the infrastructure needed to move towards a carbonless future.

Endnotes

1. United Nations, "Generating power", https://www.un.org/en/climatechange/climate-solutions/cities-pollution#:~:text=Cities%20and%20Pollution, cent%20of%20the%20Earth's%20surface.
2. United Nations, "68% of the world population projected to live in urban areas by 2050, says UN", Department of Economic and Social Affairs, United Nations, 2018, https://www.un.org/development/desa/en/news/population/2018-revision-of-world-urbanization-prospects.html.
3. Urban population in the world has grown rapidly from 751 million in 1950 to 4.2 billion in 2018. Asia, despite its relatively lower level of urbanization, is home to 54% of the world's urban population, followed by Europe and Africa with 13% each.
4. Today, the most urbanized regions include Northern America (with 82% of its population living in urban areas in 2018), Latin America and the Caribbean (81%), Europe (74%), and Oceania (68%). Urbanization in Asia is now approximately 50%. In contrast, Africa remains mostly rural, with 43% of its population living in urban areas.
5. United Nations, "68% of the world population projected to live in urban areas by 2050, says UN".

Chapter 4
ENERGY

by Sandro Desideri
and Alberto Balbo

ANALYSIS

by Sandro Desideri and Alberto Balbo

Decarbonization of the Electricity Generation and Distribution Sector

This section covers the Electricity Generation and Distribution part of energy decarbonization, following the IEA categorization of the main sectors — Electricity, Transportation, Industry, and Buildings — contributing to greenhouse gas (GHG) emissions. The latter three topics have their dedicated chapters.

2021 Climate Change: A Code Red for Humanity

A code red for humanity — this is the stark warning used by the UN Chief about the outcomes of the recently released IPCC report.[1] It is clear that humanity will not only need to implement mitigation strategies, such as shifting to renewable energy sources, with a sense of urgency never seen before, but it will also need to dramatically accelerate adaptation strategies, given that some of the changes that occurred so far, at planetary level, are irreversible on timescales of centuries to millennia. Figure 4.1 reports the scale of human influence on climate due to the greenhouse effect.

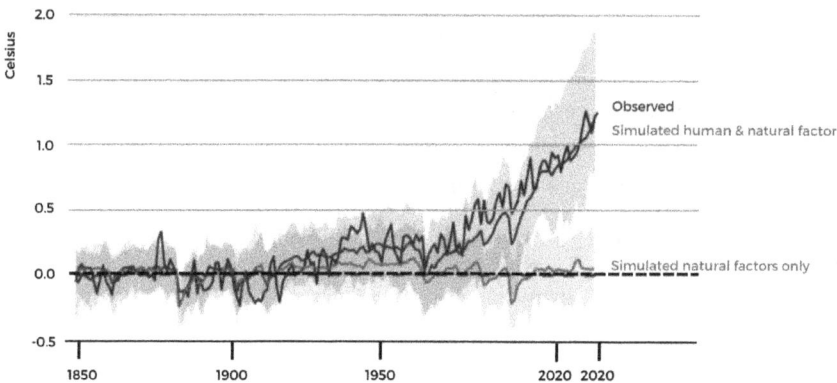

Figure 4.1. Human influence warmed the planet (1850–1920).

Source: IPCC.

Since the Paris Agreement was signed in 2015, little has changed in terms of decarbonizing human activities. The target to reduce global GHG emissions to limit global temperature increase in this century is 1.5°C. However, this objective, compared to pre-industrial levels, has not been achieved so far. Two major weaknesses have been identified: a lack of accountability and inadequate financial systems that could underpin and incentivize such a rapid transition.

According to Scott Barrett, the vice dean of Columbia University's School of International and Public Affairs,[2] the agreement fails to address the "free-rider problem" stemming from the fact that "countries would enjoy the benefits of global efforts to limit emissions regardless of their individual contribution". This creates a temptation to ride on the emissions cuts of other nations and can doom the overall effort — if everyone shirks, the global cuts never materialize. The necessity is that every single country in the world must have a target and be held accountable for it. Clear economic incentives and disincentives need to be embedded by policymakers.

When we discuss commitments and the overarching Paris framework, the most critical issue is the investments required for power systems to limit global warming. Approximately USD 3.5 trillion of annual capital investment is projected to be required between 2022 and 2050 to drive an effective energy transition. Clearly, the world is still severely lagging behind these numbers. A more sustainable financial system is slowly being built to provide funding for initiatives and innovations of the private sector and, as a result, magnify governments' climate policies. There is no time, and the window of opportunity is closing fast.

The Asia-Pacific region is currently one of the most affected areas in the world by climate change. Places such as the Pacific Islands, Philippines, and India are experiencing the full brunt of extreme weather events. The region's challenge lies in technologically leapfrogging the evolution that power systems witnessed in the West and dramatically accelerating the adoption of sustainable energy sources. At the same time, the reality is that the majority of the countries in the Asia-Pacific region are still considered developing economies that rely on dirty and subsidized energy sources such as coal. This has indeed been one of the crucial points of debate at COP26 in Glasgow, and finding a satisfactory avenue forward for everyone is one of the key challenges of our decade.

It is important to remark that only five countries in the Asia-Pacific region maintained or decreased their emission footprint in 2010: Australia, New Zealand, Brunei, Japan, and Singapore.

The Importance of the Electricity Generation Sector

The big elephant in the room is electricity generation provided via power systems. That is ultimately the sector that underpins any other, and according to certain estimates, electricity generation and distribution constitutes, with district heating and cooling (an activity strictly related to power generation in what is called "cogeneration"), approximately 50% of the planet's CO_2 emissions. Asia's fast growth trajectory is in the position of generating a predominant part of such carbon emissions, in consideration of the rapid increase in electrification rate and the parallel growth of the region's GDP. Asia, already the world's most populated region, has the biggest challenge due to the significant rise in urbanization and demographic growth.

Evolution of Energy Sources for Power Systems in Asia-Pacific

Historically, Asia has been a heavy user of fossil fuels such as coal and diesel. Targeting reliable electrification has been the challenge of the last decades in Asia, mainly in developing countries. This challenge prevented investments and opportunities for the development of renewable energy resources.

Electrification State of the Art in Asia

According to the World Bank's database,[3] the Asia-Pacific region reached an overall percentage of electrification (defined as the percentage of the population having access to electricity supply) of >98%. Countries in Asia-Pacific with still a significant gap in electrification are North Korea (49.4%), Myanmar (68.4%), and Papua New Guinea (PNG) (68.5%).

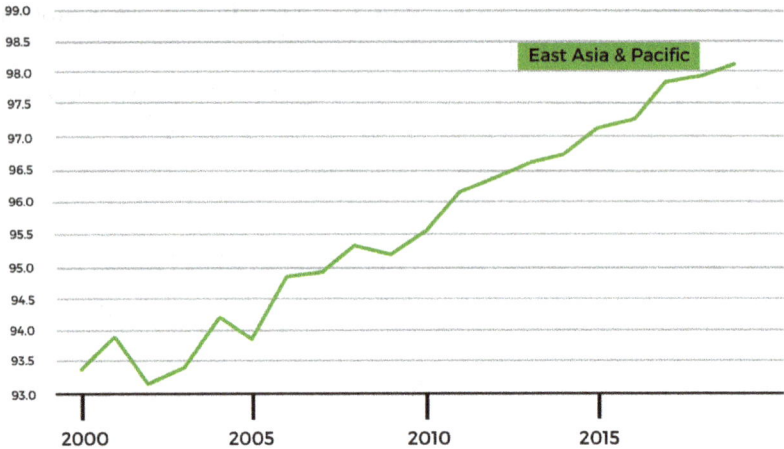

Figure 4.2.   Access to electricity in Asia-Pacific. (Y-axis: Percentage of population.)

Source: World Bank Global Database.

PNG is progressing quickly to a higher rate of electrification, whilst Myanmar is suffering from its political situation, delaying the electrification process. Figure 4.2 provides a snapshot of the great improvement in terms of electrification achieved by the region in the last 20 years.

The issue in Asia-Pacific is increasingly less about the access to electricity; it is about access to affordable, reliable, and "clean" energy. Many statistics that report dramatic data about the lack of electrification in Asia may actually be referring to the gap between the existing supply of electricity and the desired level in order to support the development of industrial and commercial activities and requirements of the social infrastructure. For example, we consider Sri Lanka, a country with 100% electrification but is now suffering from blackouts for up to 12 hours every day. There may be access to electricity, but there is not enough of it!

It is important to highlight again that a power system may be very strong in transmitting and distributing electricity but very weak in generating it or supplying fuel to the system. Other countries may need an infrastructure investment in order to deploy a suitable and smart transmission and distribution grid (T&D).

Power Systems

The generation and distribution of electricity are provided through power systems. A power system (see Figure 4.3) uses different forms of energy inputs derived from renewable sources (sun, hydro, wind, tide, biomass, geothermal, etc.), fossil sources (coal, diesel, natural gas, oil, etc.), and nuclear fuel sources, and converts them into electrical energy. Electricity gets distributed via a grid at a regional or national level or can be used for an "island" application. (In such cases, we are talking about microgrids.)[4]

Decarbonizing power systems should change the power generation of sources of energy and implement increased efficiency and loss reduction in electricity transmission. It is also paramount to consider how the grids have to be designed and managed in order to optimize the reliability, availability, and efficiency of electricity distribution. A smarter grid is a great contributor to decarbonization.

In Asia, it is also important to understand the structure of the power system industry, as deregulation has been affecting the sector over the last 30 years. Power generation is led by independent power producers (IPPs)

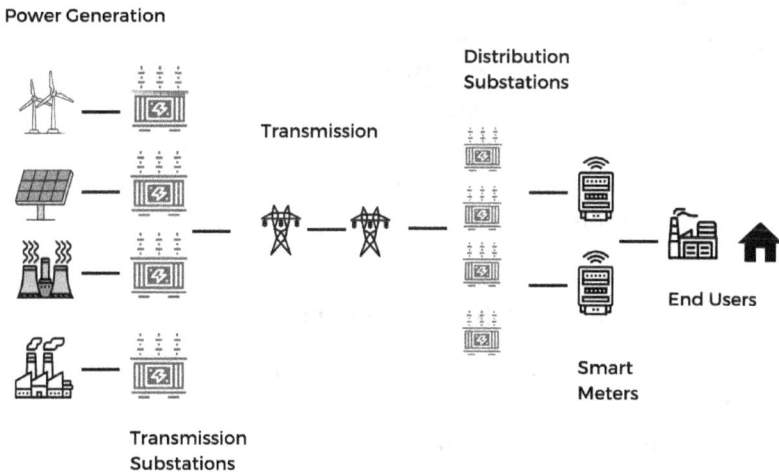

Figure 4.3. Power system schematic.

Source: Link4Success Ltd.

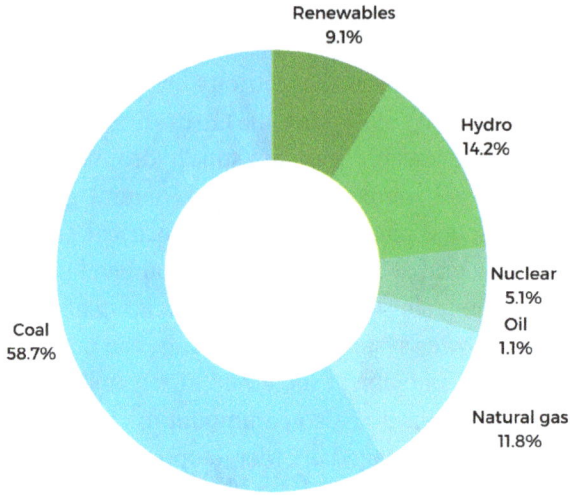

Figure 4.4. Electricity generation by fuel in the Asia-Pacific region in 2019.

that are newly created companies or spin-offs from former national utilities. Transmission is still mainly managed by the former vertical national utilities, partly for national security reasons, whilst distribution is split into distribution assets companies and retailers. Figure 4.4 reports the energy balance by source of fuel in Asia-Pacific in 2019.

The transformation has been profound and fast in all of Asia: China's power generation sector has been split across several IPPs beside the five major "gencos" — power generation companies derived from state-owned enterprises (SOEs). The transmission and distribution of electricity still remain under the control of two SOEs: China Grid and South China Grid.

Review of Major Power System Technologies

Before proceeding with the review of the new technologies that are helping to close the gap of effectiveness in decarbonization, it is paramount to get an overview of the available technologies, their actual implementation, and the challenges around them. Most of the technologies that will support the achievement of the net zero goals are already available. All that is needed is to scale those technologies after successful local testing. Testing is of the utmost importance in the deeply conservative power system sector, where the operators are tied up by big potential liabilities and are therefore very reluctant to adopt new technologies unless properly verified and validated. As an example, a new gas turbine technology is required to run more than 80,000 hours (i.e., 10 years) of smooth operations before being considered commercially acceptable.

Renewables

– As clearly reported by the IEA Net Zero report of 2021,[5] renewables are the key to archieving net zero GHG emission targets. Investments in the sector of renewables are paramount for the generation of carbon credits and, therefore, for the mechanism that will support the decarbonization challenge together and beyond the incentives from governments. However, renewables are the very visible protagonist of the decarbonization march. We will thus provide just a brief overview of them and their pros and cons in the overall context of the decarbonization challenge.

There are several types of renewables (PV solar, concentrated solar, on-shore and off-shore wind, tide, geothermal, hydro, biomass, and more). However, their most important distinction is whether they are an intermittent or a base-load (continuous) form of power generation. The most important way to rate them against fossil fuel power generation is the levelized cost of electricity (LCOE) calculation. The LCOE takes into account the cost of fuel and capital and operational investments, as well as capacity and many other factors. It gives us the cost of a kilowatt-hour (kWh) of electricity generated.

We have had amazing progress and technological innovation over the last 30 years, along with scaled-up investments in the manufacturing industry and installations around the world. All this resulted in substantial parity between PV solar, wind, and fossil power generation LCOEs. However, such parity is still presenting us with the intermittency issue, which is directly impacting the cost of transmission and distribution of electricity, as the gap between electricity demand and the effective availability of power has to be managed. PV solar, for example, is generating electricity at near-full capacity for only around 5 hours a day, for instance. Wind is very unpredictable, and even a well-located wind farm can only count on the availability of sustainable wind between 18% and 34% of the time. Intermittency drives oversizing the capacity of the plants. It also introduces the demand for energy storage as well as the hybridization of power generation by introducing fossil fuel into the mix in order to maintain the most important requirement of an electrical utility — the provision of reliable electricity.

The lack of reliability — and the challenges of managing it — has often been the major bottleneck in the shift to a fully renewable-generated power system. One way to overcome the bottleneck while staying true to a net zero approach is the integration of non-intermittent renewable sources of electricity: sustainable and certified biomass, hydro, and geothermal. Such an approach is, however, more expensive (LCOE-wise) and limited by sustainability considerations (biomass), inadequate resources (hydro), and environmental impact challenges (geothermal, hydro).

Technological innovation in the smart grid sector is the key to enabling the acceptance of a shift to more than a 60% renewable energy balance within the next decade.

It is safe to say that China is achieving regional and world dominance in the renewable energy sector, both in manufacturing capacity (see Figure 4.5) and implementation ability. China's wind and solar

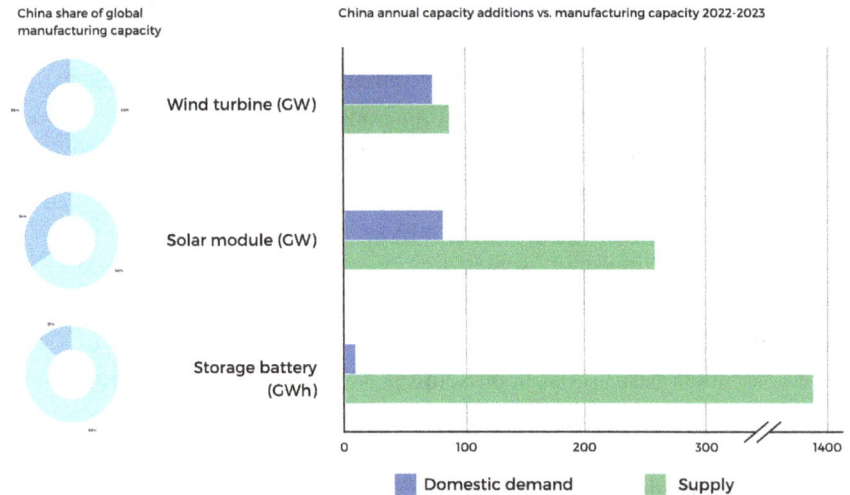

Figure 4.5.   China's share of global manufacturing capacity (left) and annual capacity additions versus manufacturing capacity (2022–2023) (right).

installations surpassed 100 GW in 2020, and it is reasonable to forecast renewable capacity additions of more than 500 GW of solar and wind by 2025 across Asia-Pacific. China announced a goal of achieving 1,200 GW of installed wind and solar capacity by 2030 to follow up on its pledge of reaching peak emissions by 2030. The excellent combination of manufacturing capacity and drive for decarbonization are very good news for supporters of the net zero goals.

It is important to mention that Japan and South Korea are at the forefront in R&D activities relating to renewables (i.e., 25 MW off-shore wind turbines) and have great ability to upgrade existing technologies. Thailand is developing a very robust biomass generation industry, and Vietnam is moving ahead in rationalizing the country's energy mix by implementing PV solar and biomass generation solutions while already generating more than a quarter of its electricity from hydro energy. However, India is still in a chaotic situation due to its inadequate implementation of PV solar and wind energy and a lack of proper management of its power grid, generating curtailment of renewable resources and an increase of coal-fired generation.

Coal

Let us start with coal-fired boiler technology, the most diffused power generation technology in Asia and the one with the highest impact in terms of CO2 emission. It is also responsible for air pollution with emissions of sulfur oxide and nitrogen oxide (these are responsible for generating acid rains), particulate matter, and many other contaminants that not only impact the quality of the air we breathe but also act as a multiplier for GHG effects!

Coal-fired power generation will eventually be phased out of the net zero pathway. The IEA World Energy Balance[6] reported a concerning situation for the Asia-Pacific region in 2019 (not considering the Oil/Diesel portion of the Energy balance, usually destined for transportation) — China and India: >80% of Energy balance dependence on Coal, Japan: ~50%, South Korea: ~45%, Indonesia: ~55%, and Australia: 60%. Thailand is an exception, with a balance of less than 30% of coal impacting their energy balance.

It is of concern that while developing countries use coal to drive their need for cheap electricity, developed countries like Japan, Taiwan, and South Korea have all increased their coal consumption in recent years. However, Japan is partially justified because of its nuclear energy exit.

The way out from coal is getting a very clear direction. Japan's Ministry of Economy, Trade and Industry is driving a policy that will prohibit the operations of non-supercritical coal-fired plants (a supercritical plant is characterized by an efficiency of ~42%) by 2025, which will impact all main power generation companies and independent power producers (IPPs). The policy will also affect the overall industrial sector in Japan, where many power generation systems are close to the end of their life cycle. This policy, together with the prospect of a heavy carbon tax, is pushing many industries to implement the blending of coal with biomass from sustainable and certified sources in order to mitigate the CO_2 non-neutral emissions. South Korea is following the same steps, resulting in a sudden rise in the demand for biomass in the region. China is the biggest investor in renewable energy over the last decade (USD 760 billion), doubling US investments made in the same period.[7]

Natural Gas

Power systems based on natural gas as the source of energy have been rapidly developing in all of Asia since the early 90s, providing a positive impact on the emissions of pollutants and particulate. The installation of large combined cycle gas turbine (CCGT) power generation plants immediately contributed to the overall efficiency of electricity generation and distribution, considering the achievable efficiency (of up to 60%) and flexibility from the operations derived by the use of natural gas. Japan has been at the forefront of such technological upgrades. This was partly the result of the abundant availability of natural gas from Australia, Indonesia, Malaysia, and the Middle East in the form of liquified natural gas (LNG).

Instrumental to such development has been the ability of the 10 major utilities in Japan (TEPCO and Chubu are among the most important) to secure a steady and certain supply of LNG by engaging gas providers such as Shell, Petronas, Exxon-Mobil, BHP, Brunei Shell, and many others, in long-term supply agreements stipulated through the trading arms of

Marubeni, Mitsui, and Sumitomo. The same strategy is recognizable in South Korea via Kepco (and all of its spin-off generation companies), Taiwan via TPC (Taiwan Power Co.), Thailand via EGAT and its many IPPs, Indonesia via PLN and its many IPPs, and, in general, the whole of Asia.

To favor the rise of the "gas-to-power", implementation has to be mentioned in the amazing scale-up of investments in gas field exploitation — both on-shore and off-shore — and in the advancement of efficiency-oriented gas turbines and gas engine technologies, which have characterized the late 90s and the first decade of the 21st century. The evolution of such an industry consisted of moving away from fossil gas to the use of renewable sources such as biogas, hydrogen, ammonia, methanol, and the use of syngas from decarbonized industrial processes. It is evident that the pathway to net zero is requiring the adoption of gas-fed power systems in order to manage the intermittency and lack of reliability of renewable sources of electricity. As reported by the IEA, the transition to 100% electricity generation requires natural gas to play a role till at least 2040.

Green Hydrogen

It is useful to provide further insights about green hydrogen (H2), the "big thing" of recent years. For some countries (Japan and Australia are at the forefront), green H2 is the stepping stone into the "hydrogen economy". Green H2 is H2 obtained by using only renewable energy while separating H2 from oxygen (O2) in the water molecule (H20) using electrolysis or other similar processes. The use of the word "green" is to distinguish the H2 generated with zero CO2 emissions from grey H2 (generated mainly by the steam reforming process[8] that generates 9 kg of CO2 for each kg of H2) and blue H2 (generated using the same process as grey H2 but with carbon capture technologies in order to reduce CO2 emissions).

Thanks to green H2, we can obtain green ammonia (power generation), green urea (fertilizer), and green methanol (transportation and power generation). Almost all the major gas turbine and gas engine manufacturers are developing and upgrading their equipment to be compatible with the use of green H2, green ammonia, and green methanol as a fuel. We already have several implementations of power generation solutions using a blending of green H2 with natural gas in the market.

However, the adoption of green H2 as a key player in the decarbonization pathway presents several challenges, of which we will mention two:

- Cost of green H2 must reach 2 USD per kg to be competitive in the market. The actual cost is 6 to 9 USD per kg, depending on the cost of the renewable energy it is generated with.
- Transportation and distribution of H2 is a challenge, considering the explosivity of H2 and its very difficult conversion to a liquid state ($-252°C$ versus $-163°C$ necessary to liquefy natural gas).

We believe that targeted investments and the scaled adoption of green H2 in complementary sectors such as transportation and fertilizers production will be instrumental to justify optimistic expectations about the net zero potential of green H2.

Companies like Hiringa from New Zealand and H2India are vowing to revolutionize how H2 is produced and, in doing so, transform the power generation, industrial, and transport sectors. In fact, it is through the ability of these companies to implement new technologies using reliable and economically sustainable solutions that an acceleration in the application of green H2 will be produced. Asia needs proof of concepts and pilots demonstrating the commercial feasibility of new solutions in order to promote scaling up implementation. The second stage will be the establishment of an industry able to manufacture equipment like electrolyzers. At this stage, Asia is a follower and not a leader, and European and US companies are ahead of the pack in technology development, with the likes of Thyssen Krupp, NEL, McPhy, Enapter, and others, setting the pace for the market.

Nuclear

Contrary to the trend in most parts of the world, many countries in Asia are planning and building new nuclear plants to cater to the growing demand for electricity. In Asia, there are about 140 nuclear power reactors in operation, with 35 more under construction. There are plans to build an additional 50 to 60. After the Fukushima "shock", nuclear is still well alive, with two-thirds of the nuclear plants worldwide to be built in Asia, and China is expected to be the main player in the sector.

However, current nuclear generation technology (fission) presents several challenges:

1) Security
2) Lack of flexibility
3) Safety
4) Very long times for the realization of projects
5) Disposal of nuclear waste

Small modular reactor (SMR) technology is answering some of the challenges — it is more flexible and takes a shorter time to install. However, (1), (3), and (5) remain very challenging; for example, SMRs are expected to play a significant role in providing at least a quarter of the baseload energy requirement of the United Kingdom. However, regulatory approval from government institutions regarding security, safety, and nuclear disposal will only be provided by 2024, and the installation of the first SMR of around 100 MW completed only in 2029!

There is also magnetic fusion technology, a potential game-changer, though there are many challenges ahead. This technology aims to reproduce what happens in our Sun, providing abundant, safe, and clean electricity — a paradigm shift for humanity.

Using the basis of today's power plant technology (fission), heavy uranium nuclei, when bombarded with neutrons, divide, and emit energy. One gram of uranium produces the same energy as 2,800 kilograms of coal. During this process, there are no GHG emissions, but the materials that remain at the end — the waste — are highly radioactive and must be disposed of and buried for hundreds of years.

Fusion, on the other hand, is the reaction that occurs in stars. At the levels of heat and pressure that form inside stellar bodies (up to 14 million degrees in the Sun), hydrogen nuclei collide and fuse into heavier helium atoms, releasing large amounts of energy, four times more than fission. To reproduce this process in the laboratory, very high temperatures (of up to 100 million degrees) are needed to fuse the isotopes of hydrogen (deuterium and tritium) and change them into "plasma", an ionized gas considered as the fourth state of matter, that is, distinct from the solids, liquids, and aeriform. This condition, then, must be maintained for a sufficiently long period of time for the final balance of the whole process to be positive, that is, for the quantity of energy obtained to exceed that introduced.

The plasma must also remain confined to a limited space without coming into contact with other structures. To obtain the result that in stars is due to the force of gravity, powerful magnetic fields are used to force the particles to follow spiral trajectories around their lines of force. The donut-shaped machine where this happens has a Russian name "Tokamak", which is the acronym for "toroidal chamber with magnetic coils". Fusion energy does not produce radioactive waste, with one small exception: the neutrons that are released make the metals they pass radioactive, but the same happens, for example, to a hospital cyclotron that produces radiopharmaceuticals. The reaction is intrinsically safe, since it goes out in seconds if the deuterium and tritium gases that feed it are not injected.

It is impossible, in theory, that this could cause geopolitical tensions, given that sufficient deuterium is found in seawater and tritium is produced in the future plant from lithium. The demand for lithium will multiply in the coming decades, but its reserves, and those of deuterium, would still be able to meet the world's energy consumption for a few billion years. The "practice", or the desire to carry out a real nuclear fusion project, was born in 1985 under the banner of international cooperation — this became known as the Iter project, in which China, Japan, India, South Korea, Russia, the United States (US), the United Kingdom, and the European Union (EU) participated. The successor to Iter will be Demo, whose task will be to pass from the demonstration of the feasibility of fusion to the production of electricity.

Energy Storage

The Asia-Pacific region is the global leader in battery energy storage system (BESS) deployment (while China is the manufacturing leader, as reported previously in the chapter), with large projects occurring in South Korea, China, and Australia. Technologically, lithium-ion continues to be the most widely used core material, as it also enjoys the spillover effect from EV technology developments. However, in the quest for true sustainability, the lifecycle of lithium-ion remains an issue, as its disposal is rather hazardous in terms of mineral extraction processing as well as environmentally challenging at the end of their life cycle. Alternative chemical compounds that guarantee 100% recyclability of BESS at the end of its life have emerged. Zinc-based and other types of so-called flow batteries (lower density batteries based on charging liquid electrolytes) are gaining ground as their cost per kWh declines. Furthermore, alternative storage technologies such as compressed-air storage, gravity storage, and pumped hydro are gaining traction. All these approaches represent incremental improvements and alternatives to the dominant BESS technology.

There are, however, even more innovative storage solutions being developed that can shake the energy industry to its very core. An example of this is the iron-air battery technology. The power in an iron-air battery comes from the interaction of iron with oxygen. The steel oxidizes almost exactly as it would during its corrosion phase within that procedure. The oxygen necessary for the reaction may be taken from the ambient air, eliminating the requirement for the cell to store it. Form Energy, a technology leader in iron-air batteries, has secured significant funding and announced a 150-MWh iron-air battery project in the US that could deliver 100+ hours of bulk storage. Such a technology, coupled with abundant renewable energy sources, could severely challenge the projected growth in consumption for power generation of natural gas in Asia-Pacific.

From Panasonic in Japan to BYD and CATL in China, well-known battery and energy storage companies in Asia that have been dominating the industry for decades, the region is full of innovation and innovators. One example is Volt14 Solutions Pte Ltd, a Singapore startup that develops ultra-high performance materials and technology for lithium-ion battery OEMs. The company that received SGD 1.3 million funding in 2021 from the likes of 500 Startups and HKSTP has developed fully integrated, tunable, and ready-to-use silicon-majority anodes. These anodes will replace current anodes, drastically increasing the performance of the batteries and reducing the costs of stationary ESSs used in grids and backup power. The other example is AMPD Energy from Hong Kong. The company designs, engineers, and manufactures state-of-the-art, grid-connected ESSs for the construction industry, a very important niche that is very difficult to "decarbonize". AMPD is now expanding, both regionally and internationally.

Virtual Power Plants

A virtual power plant (VPP) is a network of distributed power generating units, such as PV solar fields, windmills, peaker generation units, cogeneration power units, and energy storage units (batteries), working and dispatching in a connected way under the control of a VPP management system.

Thanks to VPPs, it is possible to manage the load on a grid through a smart distribution of the power generated by each of the above listed units, making it the optimal platform for trading the energy exchange. To put it simply, a VPP is a complex system to optimize the whole process of generation and distribution of electricity.

A significant amount of innovation has been delivered to the power system sector in order to implement the most effective software and IoT platforms that manage VPPs. However, we expect a rapid increase in the quality and effectiveness of advanced technologies provided to the industry in this sector.

In consideration of the geographical characteristics and actual status of implementation of the grid infrastructure, Asia is the world's region that benefits the most from VPP technologies, except for China, which invested massively in a centralized power grid and is now the leading country of the world in this sector.

A great example of a basic VPP application oriented to rural communities comes from SOLshare, a Bangladesh company that has created the first peer-to-peer (P2P) energy exchange network providing ICT-based services to underserved communities, creating synergies between electricity supply and transportation.

SOLshare came up with a way for people to turn their excess solar electricity into money with zero hassle. It also enables them to purchase more power on-the-go whenever they want and to invest in more power generation that they can trade off for a handsome return with minimal risk. This led to the installation of one of the world's first cyber-physical P2P solar-sharing grids in Bangladesh.

SOLshare has raised a total of USD 2.8 million in funding over three rounds. Their latest funding was raised on 5 July 2020 from a Venture-Series Unknown round.

SOLshare is funded by five investors, of which EDP Ventures and IIX Impact Partners are the most recent investors.

Future of Electricity Generation and Distribution in Asia

Bloomberg New Energy provided a forecast about the growth of global electricity demand until 2050, highlighting that the biggest players in the sector will be China and India, with Southeast Asia accounting for almost double the demand of the US and Europe. However, in Asia-Pacific, electricity generation from a renewable source accounts for around 20% of the whole electricity generation against ~40% of Europe.

The energy demand in the Asia-Pacific region is expected to grow from 310 to 380 quadrillion BTUs (British Thermal Units) in the period 2025–2050, with the electricity generation sector representing 266 quadrillion BTUs of such energy demand.[9] A growth of 20%! It is evident that Asia-Pacific, with its 4 billion people, is facing a very challenging task.

The roadmap of China is clear, with its huge adoption of wind, solar, and nuclear energy while transitioning coal with the help of natural gas. Southeast Asia governments are increasing their budget for hydropower implementation, with support from biomass and solar energy, while India is looking at massive wind and solar implementations to reintroduce natural gas to manage transition and instability in their power grid. India is also looking into nuclear energy for additional help.

The century of decarbonization is for sure also the century of Asia!

Conclusions

In order to get to net zero emissions, Asia-Pacific needs to accelerate its transition to renewable energy and leverage electrification and green H2 for its transportation systems while completely phasing out coal by using natural gas as a "facilitator". The shift to renewables represents a huge economic driver, and Asia-Pacific is well poised to exploit these macro trends. Governments are expected to prioritize job creation initiatives and upskilling for workers because of the significant economic decline due to COVID-19. According to the International Labour Organization,[10] up to 14.2 million global-level green jobs can be created by 2030.

If we regard renewable technologies implementation as being key for Asia-Pacific decarbonization, we need to consider the following challenges and barriers and work with institutions, governments, and private investors to remove them in the shortest possible time:

1. Cost of installation — solar and wind technologies are still far more expensive than fossil fuel-based power systems.
2. Inadequate infrastructure with limited capability to handle large amounts of renewable energy.
3. Insufficient power storage investments.
4. Non-renewable power generation monopoly — very hard to dislodge well-established practices and business models linked to the fossil fuel power generation industry.
5. Lack of knowledge and awareness about the options to achieve decarbonization while fighting climate changes.

STARTUP CASE STUDY
Hysata

by Sandro Desideri

Hysata is an Australian start-up that has developed technology and expertise for the efficient generation of green H2 from electrolysis. Green H2 is forecasted to be a trillion-dollar industry. The backbone of this industry is the electrolyzer. It is no surprise that this development takes place in Australia, considering the great attention that green H2 is attracting as a growth engine of the entire Australian economy. In fact, Australia has decided to offer itself as a large-scale renewable energy generator, with PV solar projects, concentrated solar plants, and wind farms to support the conversion of water into green H2.

Australia is committed to growing an innovative, safe, and competitive hydrogen industry as set out in its National Hydrogen Strategy. The Australian Government has since committed AUD 370 million to advance Australia's hydrogen industry. As highlighted in the relevant paragraph, the cost of green H2 depends 70% on the cost of electricity from renewable sources. Hysata is working to resolve the fundamental challenge linked to the sustainability of a hydrogen society: efficiency (Kg H2 / KWhr) of the electrolyzing process

Currently, green H2 is too expensive to compete with fossil fuels, due in large part to the low efficiencies of existing electrolyzers. Hysata's ultra-high efficiency electrolyzer is expected to make green H2 competitive years earlier than generally assumed, accelerating global decarbonization and increasing energy security.

Hysata is developing a completely new type of electrolyzer, featuring the world's most efficient electrolysis cell, coupled with a simplified balance of the plant. The Hysata founding team comprises electrolyzer industry veterans who have deep expertise in the design and scaling the technology. Backed by international investors, Hysata is moving rapidly towards manufacturing at the multi-gigawatt scale needed to address climate change in any significant way.

Hysata's technology offers step-change improvements in three key areas:

1) – New category of electrolysis with a globally leading efficiency of 95% system efficiency (41.5 kWh/kg), compared with ~75% for incumbents (52.5 kWh/kg)
   – Low-cost design, based on earth-abundant materials

2) – Simplified balance of plant (BOP)
   – High cell efficiency, which eliminates the need for expensive cooling
   – Efficient, low-cost, grid-friendly power electronics
   – Integrated BOP and stack design to provide an optimized, turn-key system that delivers high purity green H2 at the lowest levelized cost

3) – Ease of manufacturing and scaling
   – Easy to automate and scale as manufacturing is based on simple unit operations
   – Modular technology — the same basic building block for MWs to tens of GW installations.

The technology was invented by scientists at the University of Wollongong and is now being commercialized by Hysata, with backing from IP Group and the Clean Energy Finance Corporation (CEFC), and Virescent Ventures. In the 2021 seed round, the company raised USD 3.9 million, and it is now on a clear pathway to commercialize the world's most efficient electrolyzer and reach gigawatt-scale hydrogen production capacity by 2025, allowing a cost of green H2 of below USD 1.5 per kg.

CORPORATE CASE STUDY
Meralco
by Sandro Desideri

Meralco is the largest private electric distribution utility in the Philippines. Its distribution network covers ~10,000 km^2 and comprises Manila, 35 other cities, and 74 municipalities in Luzon island, serving more than 6.8 million customers. Meralco distributes ~40% of all the electricity generated in the Philippines.

Meralco procures electricity via power purchase agreements (PPAs) with independent power producers (IPPs) or on the electricity spot market. It also has more than 20 power suppliers that provide electricity out of different power generation systems — coal, natural gas, PV solar, hydro, biomass, heavy oil fuel, diesel, geothermal, and wind. It is quite a heterogenous energy mix but a significant challenge in the path for an effective decarbonization strategy.

Most parts of its power supply come from fossil fuels and only 2% from solar energy. Power generated by coal as fuel reaches 30% shares, and 15% is provided via diesel and fuel oil.

The game-changer in the net zero direction has been the implementation of the Renewable Energy (RE) Act,[11] which allows end users to become prosumers (producers and consumers), thanks to the introduction of Net Metering. With Net Metering, prosumers can produce electricity through renewable energy sources (i.e., roof solar PV installations) and use it for their own purposes while exchanging the excess with the grid and getting paid a FIT (feed-in tariff) or "credits". In 2018–2019, the net metering purchases effectuated by Meralco have grown with a compound annual growth rate of 72%. Thank to the RE Act, Meralco has been able to develop a reliable supply of ~17 MWp from photovoltaic sources, mainly solar rooftop installations.

Net Metering represents a very small percentage of the total power purchased by Meralco, but it is a very important achievement because it has helped generate a strong movement in the public for further decarbonization of the electricity distribution. The commitment from the

Center of Renewable Energy and Technology (CREST) and Greenpeace to help Meralco come out with a clear net zero GHG emission strategy has been instrumental in generating the "Decarbonizing Meralco" report.[12]

This report highlights how Meralco can shift its business focus to renewable energy while continuing to provide a reliable and affordable supply of electricity. It also defines an energy transition plan based on the six points below, to lead Meralco through the journey to net zero GHG emissions:

– All new sources of power via power supply agreements (PSAs) and other future energy demands are to be from renewable energy. No more renewals for coal supply contracts. A total of 36% renewable energy by 2030 is considered feasible.
– Lessen the reliance on an inflexible baseload and invest more in hybrid generation plants. Flexible power introduction will make renewable energy intermittency more manageable and discourage coal-based power generation investment.
– Mandate straight energy pricing for future power contracts. Meralco's practice of allowing power generation companies to pass on price fluctuation risks to their electric customers must be stopped.
– Include carve-out clauses for any future baseload PSAs. To mitigate stranded asset risks, future PSAs with inflexible baseload generators should have carve-out provisions, permitting Meralco to purchase less electricity according to changes in demand.
– Rationalize requirements for net metering.
– Pursue investments in energy storage and smart grids.

The report also highlights that the low penetration of renewables in Meralco's energy mix prevents the company and its end-users from reaping the benefits of rapid cost reductions in solar and wind energy technologies. It also shows that Meralco can reap immediate benefits from shifting power generation to renewable energy; specifically, generation costs can drop by 17%.

A scenario of 50% renewable energy in the Meralco Energy mix will bring several advantages — Meralco can enjoy a significant increase in domestic demand for decarbonized power by increasing its utilization of renewable energy. As a result, this would generate competition in the clean energy sector and drive down costs. This would help Meralco to make the right move in the energy transition roadmap.

Decreasing renewable energy costs together with the introduction of retail competition and ongoing plans to establish LNG importation infrastructure will challenge the dominance of the coal power generation. Increased coal taxation will further dampen future investments into coal and increase stranded asset risks for existing coal generators.

The first sign of the new direction that Meralco is heading is its PHP 110 billion (USD 2.1 billion) in capital spending to expand its renewable energy generation portfolio, announced on 1 March 2022. This investment is targeted towards the implementation of 1,500 MW of renewable energy generation and represents a significant step toward a coal-free 2050 energy mix. The other significant achievement is that the net metering renewable energy contribution is targeted to grow to become 10% of the Meralco energy mix by 2050. An interesting opportunity for an energy-wise complicated country to move out from fossil fuel domination!

INTERVIEW
Haukur Hardarson
Chairman and Founder,
Arctic Green Energy

by Alberto Balbo

We had the pleasure to interview Haukur Hardarson, Chairman and Founder of Arctic Green Energy, about the opportunities and challenges encountered by geothermal and other decarbonization technologies in China (and broader Asia).

Arctic Green Energy is a leading global developer and operator of profitable green energy projects. Its founding mission is to export the Icelandic success and leadership in geothermal and other renewables to the emerging markets of Asia by building up local companies. The largest company in Arctic's portfolio is Sinopec Green Energy (SGE) (www.sinopecge.com), a Joint Venture (JV) with China's Sinopec that was established in 2006. SGE has become the world's largest geothermal district heating company with 700 employees, having built 719 heat centrals across 60 cities/counties in China and drilled 724 wells. Today, the company has generated 46,500 GWth since 2006 and annually produces 15 MWe. Most of the company's activities are in Hebei, Shaanxi, Shandong, and Tianjin, and other areas are currently being added. To date, SGE has saved 16 million tons of CO_2 emissions and is expanding to other branches of geothermal such as power production and cogeneration with other green energies such as solar and wind. SGE was the first geothermal heating company to be accepted into the UN Clean Development Mechanism (carbon trading scheme).

Apart from extensive geothermal operations, Arctic's current portfolio now includes waste heat utilization and energy efficiency projects based on the Energy Management Contract (EMC) model. Profitability is a key element in building green energy companies. Arctic has shown that its ventures in Asia are profitable while at the same time it offers its customers green energy at lower prices than its fossil fuel-driven competition, without any subsidies. Arctic is a proud member of the Icelandic Geothermal Cluster and the United Nations Global Compact. Its vision of building a leading profitable green energy company is based on its core values of a cleaner and healthier world, sustainability, and social responsibility.

Considering the subject, what is the actual situation in Asia in terms of energy transition and decarbonization activities that you are observing, especially in light of the outcomes from COP26 and the 2050 pledge to net zero?

Though starting from a lower platform, Asia takes energy action and decarbonization even more seriously than most Western countries. China, for example, leads this development with its 2030/2060 declaration, i.e., a carbon peak by 2030 and carbon neutrality by 2060. Other Asian countries follow with various target dates for carbon neutrality, but the topic is being taken very seriously by the various governments.

Do you realistically think that 2050 Net Zero for China is achievable or not? Why?

Last time I checked, the target date for China was 2060, and not 2050. Having said that, I am convinced that China will reach its goals. Why? Up to 50% of the global growth in renewables is in one country, China.

What do you think it would be needed to achieve such a target?

I do think that the fastest way to decarbonize is the green energy transition of how cities are heated and cooled. Generation of heat (both space and industrial heating) typically account for over half of global energy usage, while the world's combined electricity usage is around 20% of the total. The largest component is the heating and cooling of cities, and fossil fuels still account for over 80% of the energy source for heat generation.

What do you see being the role of geothermal energy as a contributor to decarbonization?

With heating being one of the most important component of energy usage and over 80% of heat is being generated by fossil fuels, the transition of the heat source to geothermal energy has a dramatic effect. Geothermal energy turns off the chimneys immediately. Unlike high temperature geothermal, which is mostly limited to tectonic plates and volcanic areas, low temperature geothermal, ideal for heat generation, is widely available in most countries.

There is no need to generate electricity as an intermediary with accompanying energy losses. As an example, the EU today is largely dependent on gas imports for heating, and it is estimated that known geothermal resources can serve over a quarter of the EU's population with all their heating needs. The situation in China is even more attractive, with China having the largest known low temperature resources of any country. So the potential for geothermal energy to play a significant role in global decarbonization is huge — literally globally for direct use and heating/cooling, and in volcanic and tectonic plate areas for electricity generation. Then there are very interesting advances being made in Enhanced Geothermal Systems (EGS) and using geothermal re-injection as a very price competitive way of sequestering and storing carbon permanently.

What are the existing barriers that you see for geothermal energy in China and, broadly speaking, in Asia, and how do you think they could be overcome?

A typical barrier is that there is not as deep an understanding among decision-makers and creators of laws and regulations about geothermal energy. It is not as easy to understand a resource that is two kilometers underground as compared to understanding solar or wind energy. Regulations around geothermal energy seem random, e.g., legislating some under renewable or mining laws, or even water usage laws! The best way to tackle this is to continue to educate policymakers, decision-makers, and local authorities, as well as educating more engineers and scientists in geothermal energy at an university level.

What are the partners that you and Arctic would like to have in the process of decarbonizing the energy industry?

We always work with local partners in each market we are in; these tend to be energy companies. In most cases, our customers are municipal authorities, so we try to build up a strong relationship with them.

From a geothermal generation perspective, what are the technologies and innovation that you consider most attractive to play a role in the context of China's (and in a broader context, Asia) energy sector transition? Which other technologies do you see as being complementary to geothermal energy (i.e., LNG, CCHP, sustainable biomass, renewables, distributed energy generation, behind the meter solutions, green H2, etc.).

When it comes to the direct use of geothermal energy for heat generation and complementary technologies, there are exciting developments in binary technology that allow for cogeneration of geothermal heat and power at lower and lower temperatures. Ground Source Heat Pumps are also developing fast in efficiency and scale. Waste heat from industrial processes also has valuable heating potential. I am a firm believer that cities of the future will have highly correlated dual-energy systems, one carrying electricity and the other hot water for heating and cooling.

These systems will be multi-energy input (distributed and not just at a single plant) with an artificial intelligence backbone that constantly prioritizes the cleanest energy and pushes the least green energy to the back. The energy transition for any city takes years, so these systems will speed up the transition. Geothermal cogenerates well with intermittent renewables such as hydropower, solar, and wind. Carbon sequestration/storage and green hydrogen solution can be complementary with geothermal energy. While we work with LNG and biomass solutions, where applicable, LNG is, in my opinion, a transition source of energy and biomass is not sustainable.

What do you think should be the role of private investors or the private sector in the energy transition for Asia?

The private sector is driving and will continue to drive the green energy transition in Asia, and governments will increasingly focus on regulatory frameworks and incentives. I think over 80% of today's investments in renewables belong to the private sector.

By converse, what role do you see multilateral institutions such as the World Bank, Asian Development Bank, International Finance Corporation, and non-governmental organizations playing in the energy transition?

These organizations entities typically declare high ideals, but the reality is often different and terribly bureaucratic and cumbersome. I think the clean energy transition will do fine with or without them.

Do you and Arctic have a model (country and/or company), if any, for sustainable development that you would like to emulate?

We are clearly built on the foundations of the Icelandic geothermal energy resources, with a cascading usage of geothermal energy at different temperatures. Iceland was the world's first country to generate 100% of its primary energy from renewables. To that can be added the Chinese model we have created through our JV in China, Sinopec Green Energy, which is today the world's largest and fastest-growing geothermal district heating company.

With regard to COP26, how would you rate the event in terms of its ability to positively influence international cooperation in order to achieve net zero within 2050 (or earlier)?

I think COP26 was very successful in positively influencing international cooperation and generally pushing things along. Of course, it can be argued that things are never fast enough, but things are definitely heading in the right direction. I have particularly high hopes for COP28 in Abu Dhabi. It is a sign of the times that a country whose wealth is based on fossil fuels is hosting COP.

Being very direct and blunt, who or what do you consider the worst enemy (or enemies) for the energy transition in Asia. Conversely, who or what are the best friends and allies?

I think central governments and the public agree on energy transition. No one likes to see their elders or children getting sick or the environmental devastation that comes with using fossil fuels to drive societies. The opposition is, of course, the competing industries and the fossil fuel lobby that knows its days are coming to an end but yet try hard to squeeze in more time for themselves. However, more and more fossil fuel interest are putting their might behind the green energy transition, such as our partners in China, Sinopec.

Any other relevant point that you would like to remark on about energy transition in the context of Arctic and its operations? How we heat and cool our cities is the single-largest energy challenge.

Geothermal energy has been proven on a massive scale to be the most effective weapon for decarbonization. We are currently working on over 60 cities and towns. We take down the chimneys and help sequester some of the fugitive carbon already in the air back into the ground where it belongs, using the same infrastructure.

Endnotes

1. IPPC (Intergovernmental Panel on Climate Change — 2021 Report) Intergovernmental Panel on Climate Change, *Climate Change 2021: The Physical Science Basis. Contribution of Working Group I to the Sixth Assessment Report of the Intergovernmental Panel on Climate Change.* Cambridge, United Kingdom and New York: Cambridge University Press.

2. Nicholas Kusnetz, "Why the Paris Climate Agreement might be doomed to fail", *Inside Climate News*, 28 July 2021, https://insideclimatenews.org/news/28072021/pairs-agreement-success-failure/.

3. The World Bank, "Access to electricity (% of population) — East Asia & Pacific", https://data.worldbank.org/indicator/EG.ELC.ACCS.ZS?locations=Z4.

4. The basic elements of an electrical grid downstream the power systems are, in sequence, step up transformers and substations (to adapt the voltage and manage the transmission line requirements), transmission lines (from 11 kV to 1,200 kV), step down transformers and distribution substations, distribution lines, and meters (to measure and manage smart uses of electricity).

5. International Energy Agency, "Total energy supply, 2019", https://www.iea.org/regions/asia-pacific.

6. Statista, "Forecasted energy demand in Asia Pacific Region", Statista, 2022.

7. The Center for Renewable Energy and Sustainable Technology, "Decarbonizing Meralco: The imperative to prioritize pro-people, pro-climate models in the power business", Greenpeace Southeast Asia, April 2021, https://www.greenpeace.org/static/planet4-philippines-stateless/2021/04/32e45a82-greenpeace-decarbonizing-meralco-04-14-final.pdf.

8. Steam methane reformation process: In steam-methane reforming, methane reacts with steam under 3–25 bar pressure (1 bar = 14.5 psi) in the presence of a catalyst to produce hydrogen, carbon monoxide, and a relatively small amount of carbon dioxide. Steam reforming is endothermic, meaning that heat must be supplied to the process for the reaction to proceed.

9. Statista, "Forecasted energy demand in Asia Pacific Region".

10. International Labour Organization, "World employment social outlook, trends 2018", International Labour Organization, 2018, https://www.ilo.org/wcmsp5/groups/public/—dgreports/—dcomm/—publ/documents/publication/wcms_615594.pdf.

11. Ranulfo Ocampo, "How net-metering works: Understanding the basics of policy, regulation and standards", Department of Energy, Republic of

the Philippines, https://www.doe.gov.ph/1-how-net-metering-works-understanding-basics-policy-regulation-and-standards.

12. The Center for Renewable Energy and Sustainable Technology, "Decarbonizing Meralco: The imperative to prioritize pro-people, pro-climate models in the power business".

Chapter 5
BUILT ENVIRONMENT

by Eric Chong

ANALYSIS

by Eric Chong

As the world endeavors to scale up its response to limit global warming, cities remain at the forefront of the battle against climate change. The reason is clear — whilst they take up less than 3% of the world's land area, they use up almost 78% of its energy, accounting for more than 60% of the global greenhouse gas (GHG) emissions.[1] In this regard, cities in Asia are in the vanguard, given that the region accounts for the majority of GHG emissions on this planet (see Figure 5.1).[2]

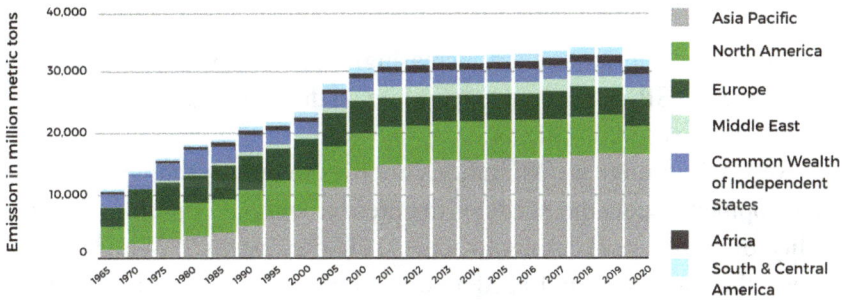

Figure 5.1. Carbon dioxide emissions by region.

The Built Environment and Its Impact on Sustainable Development in Asia

More than half of the world's population now lives in urban areas, a proportion that is expected to increase to 68% by 2050 (see Figure 5.2).[3] A total of 754 million of these urban dwellers reside in Asia-Pacific, making up 61% of its total population that contributes to over 80% of its GDP.[4] The rapid urbanization and consequent proliferation of megacities[5] across the region comes with a price. Expanding cities consume already-depleted resources to the detriment of their biodiversity and their inhabitants. Cities in Asia are thus pivoting to sustainability as a credo in their

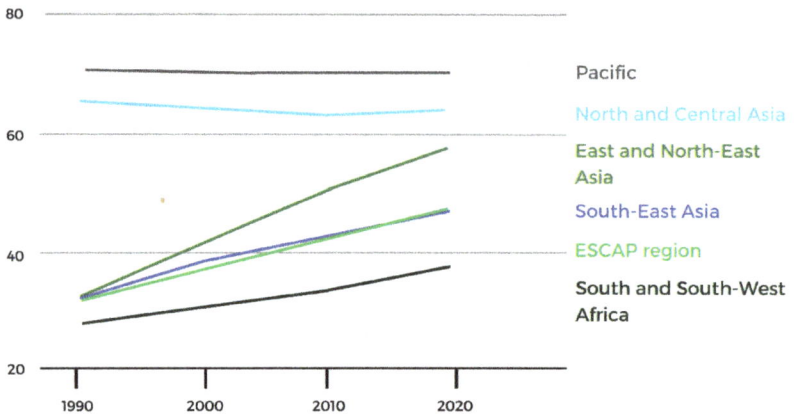

Figure 5.2. Urbanization in Asia and Pacific regions.

development blueprint and are increasingly embracing "sustainable built environment" (SBE) as a core tenet. More on this in the interview with Dr. Cary Chan later in this chapter.

An SBE captures the aspirations of cities to grow sustainably so that "development meets the needs of the present without compromising the ability of future generations to meet their own needs."[6] Whilst that is indeed the aspiration and academic wisdom, SBE in practice is easier said than done. It firstly needs cities to completely rethink how resources are used in their built environment and adopt a circular economy: reduce, reuse, and recycle must become part of daily life. It also needs an embrace of renewables in place of non-renewable energy to reduce carbon use and emissions in the long term. Most of all, it means recognizing the natural costs to the environment and biodiversity of its development decisions. This complexity is exacerbated by the multitude of stakeholders that need to be considered. Regardless of whether they are social, political, economic, or legal, they can affect its realization either positively or negatively.

Since the launch of the Sustainable Development Goals in 2015, attention has once again been brought to how the SBE impacts sustainable development in general.[7] This reminds everyone that the SBE does not exist in a vacuum. It is highly contingent on different levels of jurisdiction, the environmental constraints, and their interconnection. Its development depends on a thriving, sustainable construction industry as it is founded on environmentally sustainable building practices. Sustainable communities benefit from it through its resource-efficient and environmentally friendly characteristics. An SBE, together with its different levels of response, thus plays a key role in global sustainability (see Figure 5.3).[8]

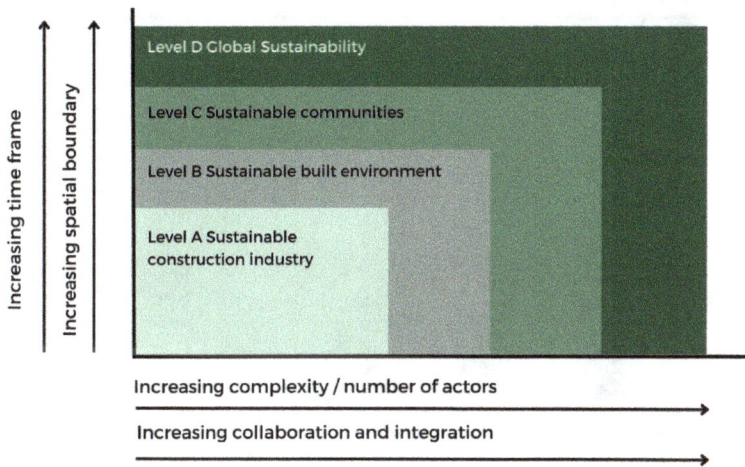

Figure 5.3. Levels of response to sustainable development.

PAST

"Green buildings" in Asia started more than 30 years ago when Hong Kong launched the "Hong Kong Building Environmental Assessment Method" (HK-BEAM). Hong Kong's induction came on the heels of the United Kingdom's passage of its building environmental performance assessment system developed by its Building Research Establishment (BRE). Known as the BRE Environmental Assessment Method (BREEAM), it was a pioneer in the industry that spawned many efforts by

Launched	Country/City	Certification	Enforcement	Buildings certified (by the end of 2011)
1990	U.K.	BREEAM	Voluntary/Compulsory for government buildings and schools	Over 200,000
1996	Hong Kong	BEAM/BEAM Plus (since 2009)	Voluntary	Over 200
1998	U.S.	LEED	Voluntary/Compulsory for public buildings in some states	Over 40,000
1999	Taiwan	EEWH	Voluntary/Mandatory for public sector (central government)	Over 500
2001	Japan	CASBEE	Voluntary/13 local governments in Japan have made it mandatory for certain businesses	Over 190
2003	Australia	Green Star	Voluntary/Mandatory for all commercial buildings more than 2,000 square meters	Over 250
2005	Singapore	Green Mark	Voluntary/Regulations require all new buildings be constructed to the Green Mark standard since April 2008	Over 500
2007	China	GBL	Voluntary	Over 200

Figure 5.4.   Green Building schemes in Asia.

jurisdictions around the world, including in Asia, that have rolled out their own certification schemes (see Figure 5.4).[9]

Since their launch, green buildings, also referred to as "sustainable buildings", "high-performance buildings", or "green construction", have gained recognition from governments and industry.[10] They involve resource-efficient building, renovation, operation, maintenance, and demolition methods. Developers of green buildings adopt a holistic view of buildings by considering their entire life cycle, from cradle to grave, starting at the planning, design, and construction stages to the building's operation, renovation, and the final fate of its materials.[11] A notable example in Asia is Shanghai Tower, the world's highest green building and the second-tallest skyscraper in the world at a height of 632 m. Completed in 2015, it achieved China's three-star green building status and is certified gold according to the United States' LEED rating system. It was built with more than 43 sustainable features, including a unique glass envelope that insulates the building from external heat and cold while providing natural light. Overall, the Shanghai Tower achieved a 21% reduction in its carbon footprint, saving 37,000 tons of carbon annually compared with buildings of similar size.[12]

Green buildings have brought a higher awareness of sustainability to the built sector. This is critical to decarbonization because buildings are major users of electricity in cities. In Hong Kong, where buildings consumed up to 89% of its electricity, the government has promulgated policies to drive energy efficiency as a key part of its "demand side" decarbonization strategy.[13] This is critical because whilst large investments are made to reduce carbon emissions on the "supply side" — generation of zero-carbon electricity — equal attention needs to be paid to the "demand side". This will prevent the decarbonization effort from becoming a Sisyphean task where increasing energy consumption such as by buildings wipes out the carbon reductions realized in the electricity supply.

The focus on sustainability and decarbonizing of the built sector has resulted in the growth of green buildings. Given the existing performance gaps, this offers tremendous opportunities to leverage innovation to further improve the energy efficiency of buildings as well as drive advancement in related fields such as smart grids and renewable energies (see Figure 5.5).[14]

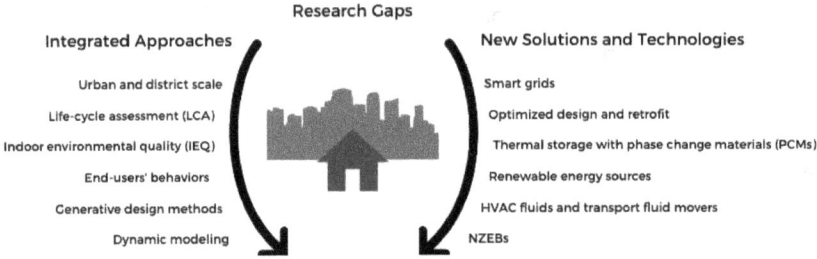

Research Gaps

Integrated Approaches	New Solutions and Technologies
Urban and district scale	Smart grids
Life-cycle assessment (LCA)	Optimized design and retrofit
Indoor environmental quality (IEQ)	Thermal storage with phase change materials (PCMs)
End-users' behaviors	Renewable energy sources
Generative design methods	HVAC fluids and transport fluid movers
Dynamic modeling	NZEBs

Figure 5.5. Opportunities for improving energy and environmental performance of buildings.

In recent decades, innovation in "building management systems" (BMS), aided by the increased energy efficiency of plants and equipment, has somewhat helped narrow the performance gaps. Unfortunately, gains from such systems, which automate the control of a building through a "command and control" scheme, seem to be approaching a point of diminishing return, limited by the available technology. This has given rise to research on smart buildings, a concept that uses system automation, computer-assisted management, and active controls to optimize a building's environmental performance.

PRESENT

Although interest in smart buildings commenced in the 1980s, these were far from the representation of smart buildings that scholars envisaged. Academics argued that such buildings should be "intelligent" enough to respond to the needs and well-being of its occupants while being functional and sustainable enough to contribute to the environmental needs of society.[15] Technological limitations during that time, however, meant that actual implementations were rather rudimentary. The main challenge arose from the dispersed nature of a building's often inaccessible parts, such as the equipment room, utility channels, and rooftops, compounded by the low permeability of its structure. This made connecting a building system to collect data, the bedrock of a smart building, a conundrum for engineers.

This situation has changed in recent years because of rising investments in technologies such as the "Internet of Things" (IoT). This has enabled the collection of massive amounts of data from connected devices at a scale hitherto impossible. This development makes possible "smart infrastructure", defined as "systems that use a feedback loop of data, which provides evidence for informed decision making".[16] Smart infrastructure will be indispensable to the global decarbonization efforts as cities scale up their efforts to drive energy and environmental performance.

Whilst perceived as a recent phenomenon, IoT had its beginnings more than 30 years ago when a Trojan Room coffee pot was connected to the internet. That was quickly followed by a toaster that could be turned on and off — the first known instance of a remotely controlled web device.[17] These first instances, just after the introduction of the World Wide Web, gave birth to the idea of connecting devices to internet networks. This interlinked web of devices allowed distributed communication between hardware, with limited interaction by humans. IoT remained very much a concept in the laboratories until 2000, when Asia electronics giant LG released its smart refrigerator. The IoT refrigerator kept track of food stored to enable faster restocking.

Since its introduction to consumers two decades ago, IoT has been increasingly embraced by commercial companies. The launch of the IPv6 standard in 2011 allowed millions more devices to connect to the internet, previously not possible with the limited scale of IPv4 addresses. Companies like Cisco, IBM, Ericsson, and Huawei took the lead in bringing new generation IoT devices to market. This research-industry cooperation led to the rapid growth in the adoption of IoT devices for infrastructure.

The development in IoT has been spurred by rapid advances in hardware.

First, progress in sensor technology made IoT devices smaller, less expensive, and extended their battery life. An excellent example is "IoT Pixels",[18] which are low-cost tags the size of a postage stamp that could be attached to any item to collect data. Their uniqueness lies not only in their size but their ability to transform radio frequencies into viable power, creating continuously running self-sensing devices at a scale previously unattainable.

Second, advances in broadband technology created a robust communication network with high bandwidth, accommodating the colossal amount of devices envisioned in smart infrastructure. Whilst "Low Power Wide Area Networks" (LPWAN) technologies such as "NB-IoT" are already deployed in smart buildings, it is the fast-paced growth of 5G since 2018 that is creating excitement in the industry. Asia, and especially China, has taken the lead, both in technology development and adoption. Though five Asian and European companies owned more than 80% of inventions connected to 5G standards, Chinese IT giant Huawei dominates by having the most patents on next-generation 5G technology.[19] Asia similarly reigns supreme in deployment. South Korea leads with 5G rolled out in 85 cities by January 2020, followed closely by China with deployment in 57 cities.[20] 5G will have a profound impact on smart infrastructure, given its extreme speed, ultra-low latency, and massive bandwidth available over a much wider area than current LPWAN technologies. This positions it to be the new industry standard for smart infrastructure and smart city development.

Third, IoT has benefited greatly from innovations in artificial intelligence (AI). If sensors are the eyes and ears of IoT and 5G is its nervous system, AI is its brain. The massive investment in AI, founded on big data analysis and machine learning, has resulted in a positive spillover for IoT. New software can capture, learn, and simulate the role of human management. Such intelligent systems can operate and optimize infrastructure autonomously. In recent years, this advance has been enhanced by the arrival of edge computing, a concept where computing capabilities are built into the IoT device themselves. This frees up bandwidth, enhances data security, and speeds up response time since the device does not need to rely on a distant server for its computing power. An example of an edge application is in image recognition, where instead of transmitting the whole image, the edge device analyzes and codes it internally, sending only this code to the cloud, thus reducing the amount of data to be transmitted.

The above technological innovations have had a profound impact on the digitalization of infrastructure. Engineers are now able to supervise, control, simulate, and predict the operation of their devices in the "Cyber Physical space",[21] commonly referred to as a digital twin. These capabilities allow operators to intelligently optimize a building's performance at all times and at a higher precision than was possible before. Their most immediate deployment has been in improving the efficiency of the "heat, ventilation, and air conditioning" (HVAC) systems, since these account for up to 50% of a building's energy consumption.

Applying smart infrastructure technology to reduce energy consumption has spawned particular interest from investors in Asia, where energy demand in commercial buildings is projected to triple by 2040. This has forged a growing market for energy-efficient products and solutions, estimated to reach USD 17.8 trillion by 2030 in East Asia Pacific and South Asia.[22] An excellent example is Barghest Building Performance in Singapore, which promises up to 40% reduction of energy costs with zero upfront investment. Detailed information about this application is explained in the startup case study.

FUTURE

As innovation in the smart infrastructure space makes buildings more intelligent, its broad mission to "enhance the quality of life for its occupants, advance human activities"[15] remains unaltered. In this regard, the introduction of "Net Zero Energy Buildings" (NZEB),[23] the North Star for Sustainable Built Environment, will drive its future as it metamorphosizes to further cement its position in the SBE.

Commonly referred to as "Zero Carbon Buildings" in Asia, examples such as the aptly named ZCB in Hong Kong offer a glimpse of how buildings will further transform to fulfill its vision of contributing positively to the ecosystem through "balancing the use of passive, active, or hybrid strategies to meet both human and ecological needs".[24] They present an opportunity for the built environment sector to realize its aspirations to truly contribute to the global fight against climate change. The zero-carbon building movement will transform the sector and accelerate innovation in its effort to reduce emissions in all aspects of the building's life cycle. More critically, it aims to move beyond the current focus on energy efficiency by reducing carbon use in operations with zero-carbon energy. The development of zero-carbon buildings will be driven by upcoming innovations such as:

Green concrete — A term that is keeping investors excited these days. There is a good reason for this. Next to emissions produced during the operation of the building, i.e., operational carbon, embodied carbon contributes 49% of the total carbon emissions of a new construction. The majority of embodied carbon is from cement needed to make concrete, an essential building block of buildings, especially in Asian cities.

Whilst the industry is strenuously working to reduce emissions from cement production, which accounts for 7% of the world's CO2 emissions,[25] it is looking at substituting it with innovative low carbon "supplementary cementitious materials" (SCM). These are a class of materials that have similar properties to cement when exposed to water but are derived from "pozzolans" — non-fossil or recycled sources.[26] For instance, recycled glass can be grounded to produce glass pozzolan, which can be added to the concrete mix to reduce the amount of cement needed. An excellent example from Asia is a pozzolan comprising ash and rice husks, a common by-product of agriculture in the region.[27] The use of SCM can lower the overall need for Portland cement in the concrete by up to 40%, creating "green concrete" indispensable in zero carbon buildings.[28] For instance, Hong Kong-based Gammon Construction's use of "Gammon Green Concrete" has helped it to reduce almost 20% of its cement use in 2020, resulting in an estimated 23,000 ton reduction in embodied carbon.[29]

Green skyscrapers — these represent a more revolutionary approach to reducing embodied carbon. They replace concrete completely with a more sustainable and natural material like timber. The use of new techniques such as "cross-laminated timber" (CLT) is making it possible for architects and engineers to design skyscrapers out of timber. This trend is poised to literally shape the skyline, with projects developed and in planning in various cities in Europe and North America.[30]

In Singapore, the Nanyang Technological Institute is building the largest wooden building in Asia in support of the government's 2030 Sustainable Development Agenda.[31] Timber's green credentials come not only from its substitution of concrete but also from its source: trees that absorb carbon dioxide during their lifetime. Locking that carbon dioxide in timber used in wooden skyscrapers offers a convenient depository for sequestered carbon, which otherwise is released into the atmosphere if the tree dies in nature.

PV facade — another area to watch as photovoltaic (PV) systems to generate solar power are a popular route for buildings in their quest to lower emissions.

The solar PV industry has grown 35% in the last decade, mainly in Asia, particularly China, driven by a combination of improved technology, reduced cost as well as policy incentives. Whilst the conversion efficiency of PV panels languished at below 20% for many years, innovation in the last ten years has improved this to approximately 22–23% in recent years. This increase, coupled with the scaling up of production, has driven down the cost of solar energy, whereby a levelized cost of energy (LCOE) for utility-scale PV at USD 32 to 44 per MWh is lower than a natural gas-fueled combined-cycle plant.[32] However, this level of cost competitiveness may be difficult to achieve in cities that are challenged by land constraints where the main solar capacity is generated from PVs deployed on rooftops of high-rise buildings.

Building-applied photovoltaics (BAPV) solve this problem by integrating PV cells in the vertical façades, multiplying the area for harvesting solar energy. It is estimated that if Singapore adopts such technologies, it can potentially generate an additional 1.7 GWp, almost doubling its target of generating 2.0 GWp of solar energy by 2030. Towards this, it is already making investments in lighthouse projects such as PSA Tuas Port Maintenance Base's administrative building, which uses solar cladding instead of aluminum cladding.[33]

Passive homes — whilst not new, is a technology that takes advantage of architectural design to maintain a comfortable temperature inside buildings. This addresses a core tenet of sustainability in Asia — minimizing the amount of energy used for cooling buildings since most cities in the region's tropical areas depend heavily on air-conditioning. Singapore is a typical example where Mr. Lee Kuan Yew, the country's late founding father, once commented that air-conditioning was the "greatest invention of the 20th century". The dependence on artificial cooling presents a vicious cycle, where the energy supply is drained to cool the interior, simultaneously raising external temperatures through its exhaust, which aggravates the heat island effect. This is when urban infrastructure absorbs and re-emits the heat from the Sun, leading to relatively high temperatures compared to natural landscapes.[34] This untenable situation, exacerbated by global warming, will mean that energy use on air-conditioning may grow by 73% by 2030.[35]

Passive homes are "designing for the local climate" and take advantage of natural climatic features such as sunlight or wind to effectively reduce the need for energy-guzzling air-conditioning.[36] These are often accentuated with elements, such as insulation, glazing, windows, and skylights, to increase the overall comfort for its occupants. Its growing adoption in Asia,[37] with examples such as the "SDE4" net-zero energy building at the National University of Singapore, illustrates its escalating importance in decarbonizing the region.

The technologies described above certainly pave the way for further developments in the zero-carbon buildings space. Whilst its future appeared to be centered on the important goal of reducing energy and carbon use, it is also crucial to remember that an SBE must holistically consider all natural resources used in the built environment. The LEED Zero program launched in 2019 reminds us of this with its aim to recognize buildings that have achieved net-zero goals in carbon, energy, water, and waste.[38] Water, in particular, deserves special attention, given its elemental importance to cities, both for the survival of its inhabitants as well as for its effective functioning. Threats like "Day Zero" at Cape Town in 2017 has brought an increasing awareness of the need to build water resilience in cities.[39] The challenge from increasing demand brought about by population growth, coupled with decreasing water sources often caused by climate change, promises to bring this critical resource more to the fore in the future.

Leadership: The Key to The Future of Sustainable Built Environment

In her keynote speech at the 17th World Sustainable Built Conference held in Hong Kong in 2017, Christina Figueres, who is recognized as the architect of the Paris Agreement, envisioned a future where cities will be crowded but clean and connected. To achieve this, she placed the onus on the sector to "innovate and integrate", as the built environment today will decide the quality of life in the future.

Leaders in SBE thus have a huge responsibility to ensure that it lives up to its aspiration as "a process of change in which exploitation of resources, the direction of investments, the orientation of technological developments, and institutional change are all in harmony".[40] One key role will be to drive and scale up innovation crucial to its future.[41] For a start, leaders in government must take the lead to establish a framework that balances regulation with support for innovation.[42] At the same time, leaders in the private sector will need to step up and take action, given that they have the unique wherewithal to resource innovators and drive adoption. To make a bigger impact on decarbonization, the public and private sectors should work together to scale up their efforts. EXPO2020 in Dubai is an excellent example where the government and German technology leader Siemens are collaborating to showcase how innovation can drive decarbonization (see the corporate case study).

As delegates departed from the 2021 COP26 conference, there was overall disappointment at how little progress the world has made since the landmark Paris Agreement in 2015. Whilst nearly 70% of the COP21 signatories, accounting for 57% of global emissions, submitted new or updated "nationally determined commitments" (NDCs), these, however, make a modest dent in the global drive to decarbonize.[43] More leadership will be required from governments and the private sector to accelerate the world's drive to reach carbon neutrality by mid-century, mitigate the effects of global warming, and avert a long-term climate crisis. Doing anything less may mean that the world will miss a key bulwark against the existential consequences of global climate change.

STARTUP CASE STUDY
Barghest Building Performance

with Eric Chong

The Future of Data Driven Sustainability

Barghest Building Performance (BBP) is a Singapore-based startup founded in 2012. It leverages IoT, proprietary software algorithms, and AI to help businesses achieve energy and cost savings. Its patented HVAC optimization technologies are focused on cooling, which is responsible for up to 60% of the energy consumption in a building. The startup's business model requires zero capital outlay by building owners.

BBP's founding goal was to help businesses decarbonize with the least number of barriers possible. Accordingly, the company funds all up-front investments and implementation to achieve energy savings for industrial and commercial buildings. Building owners engaging BBP share a portion of their actual cost savings with the startup. These savings are independently verified by third-party auditors.

In the past decade, BBP has implemented its solutions across the Asia-Pacific region in semiconductors plants, heavy industry, data centers, pharmaceutical companies, cold chains, and commercial real estate. So far, they have achieved up to 40% energy and cost savings without any capital investments or outlays from the companies. Since its founding, the startup has enabled the avoidance of 100 million kilograms of CO2e emission and saved more than SGD 26 million (USD 19.1 million).

Holistic Process to Decarbonization

BBP addresses the typical pain points of property owners and operators, such as aging equipment, faulty parts from poor maintenance, and unoptimized energy management through a data-driven approach.

The first step in deploying the technology is a detailed onsite audit and analysis. The BBP team surveys the building owners to understand their operational constraints before making recommendations for optimization and control strategies.

Post-implementation, the company continues to monitor customers' chiller plants and equipment remotely to detect any drift in performance. Such real-time insights enable long-term collaboration on preventive maintenance, delivering annuity income for the startup whilst providing additional savings on operational costs for the customer.

According to BBP, their main differentiation is addressing both internal and external environmental conditions at every building, helping customers achieve the best possible optimization outcome without compromising operations.

Making Buildings Smarter

Since its founding, BBP has placed great emphasis on developing holistic data analytics and AI. In 2020, it launched an advanced analytics platform that offers fault detection and diagnosis, reducing false alarms and operational anomalies, identifying system errors in building HVAC systems, and managing the equipment's life cycle. This offering is symbolic in the industry as demands of smart buildings shift from reactive problem solving to proactive maintenance. As experts repeatedly point out, the primary goal of achieving gold standard efficiency is fading away, and buildings are looking for solutions and platforms to anticipate real-time impact and risks.

In the last decade, BBP has delivered solutions to highly complex and stringent operational environments like semiconductor facilities, pharmaceutical plants, data centers, and cold chains, all of which run 24/7. As of 2021, BBP serves customers across Southeast Asia, China, India, and Taiwan, including three of the world's top 10 semiconductor manufacturers, four Fortune 500 companies, and some of Asia's largest real estate companies.

Some notable projects include:

Semiconductor Property — Lumileds Manufacturing Plant

The Lumileds Singapore plant, which operates round-the-clock, comprises two blocks and has a total Gross Floor Area of 41,716 m^2.

Without replacing its 14-year-old chiller plant or disrupting operations at the plant, BBP helped Lumileds achieve sustained 30% savings of initial energy consumption, saving SGD 700,000 (USD 512,000) in annual energy costs. This plant (see Figure 5.6) is the first manufacturing plant in Singapore to receive the Green Mark Platinum Award.

Figure 5.6.   Lumileds Singapore Plant.

Commercial Property — Resort World at Sentosa, District Cooling Plant

The district cooling plant supplies chilled water to the entire property, including the casino, hotels, and Universal Studios Singapore (see Figure 5.7).

Figure 5.7.   Resorts World Sentosa.

BBP uses an algorithm-based control to achieve and maintain better chiller plant efficiency across varied usage without affecting the comfort of occupants and visitors. The successful implementation of the optimization project resulted in a 10% improvement in energy efficiency, saving 4 GWh annually.

Industrial Property — Hewlett-Packard Production Facility

HP Inc. (see Figure 5.8) worked with BBP to create customized solutions for its chilled water systems, which had been in operation for more than 20 years.

The partnership delivered an average of 22% energy savings annually for HP, totaling 2.5 GWh since implementation.

Figure 5.8. Singapore HP data center.

The Future of Data-driven Sustainability

The energy savings delivered by BBP's optimization efforts translate into carbon dioxide emission avoidance, enabling building owners and operators to reduce their overall carbon footprint using data. These data-driven solutions promise to form the core of the world's effort to reach carbon neutrality. They will also translate into new business for those innovative enough to comprehend the opportunities. For instance, BBP is looking into helping its customers generate carbon credits using such data to capitalize on the growing market for carbon offsets.

Since its founding in 2012, BBP has attracted investments from the capital market. In December 2018, it raised SGD 45 million (USD 32.8 million) from Kohlberg Kravis Roberts and others in Series B funding. Today, BBP has expanded its business in Asia-Pacific, deploying its solutions in nine markets, with the latest being Vietnam.

CORPORATE CASE STUDY
Expo 2020 Dubai

with Eric Chong

A Blueprint for Impactful Collaboration

For Siemens, a global technology company founded in 1847, sustainability is a business imperative. Employing approximately 300,000 persons worldwide, the company serves diverse customers in more than 130 countries, focusing on industry, infrastructure, transport, and healthcare.

Just before COP21 in Paris, Siemens announced that it would make its global operations carbon neutral by 2030. Whilst this reflects its commitment to sustainability, it also sees this as an opportunity for growth. It has accordingly developed an integrated "Environmental Portfolio" related to carbon reduction that includes products, systems, solutions, and services that support its customers' decarbonization efforts. For FY 2021, 87.5 Mt of GHG were reduced through Environmental Portfolio products in operation by such customers.[44]

Siemens expects the decarbonization of Asia to reshape industries and countries across the region. The company is thus taking steps to tap the opportunities presented and further contributing to the battle against climate change. The collaboration[45] with EXPO2020 in Dubai to demonstrate the future of sustainability in cities is one such example.

Dubai, as the host city for EXPO2020, offers both challenges and opportunities in sustainable development no different from those faced by other cities in Asia. With a population of 3.4 million and a land area of 35 km^2, increasing constraints are being placed on its limited natural resources. Whilst rapid urbanization over the last three decades has resulted in profound economic growth, it has unfortunately also brought major concerns. Rapid urbanization in Dubai is being blamed for many environmental and social problems, most of which require long-term solutions.

Innovation is seen as an answer that can help Dubai "do more with less", making a long-term impact on productivity and sustainability. For instance, investments in smart infrastructure could make more efficient use of resources and enhance the way people live, work, and play. Its highly automated and intelligent features may even help the city address its reliance on non-nationals in its labor force.

The deployment of smart infrastructure technologies can benefit a city like Dubai in multiple ways. Rather than a single building reducing its energy consumption, citywide solutions with smart programs that dim street lights whenever there is low traffic scale up those savings exponentially. Such technologies also make practical the implementation of smart grids to manage the increasing percentage of renewable energy in the electricity mix whilst allowing consumers to play a bigger role in energy saving through demand-response schemes.[46] The impact of smart technologies does not stop at energy, as the same principles can be further extended to manage other scarce resources such as water and air.

Siemens' partnership with the city aims to demonstrate and showcase how these technologies can support urban sustainability and involve building an eco-friendly urban environment at the Expo site to serve as "a blueprint for future smart cities". Lighthouse projects in the buildings, energy, and water domain include the following.

Smart buildings

Buildings are responsible for 40% of global energy consumption. It is also the place where most people spend their lives. Siemens uses its "Navigator", a cloud-based energy management platform, to connect, collect, and analyze environmental and operational data from thousands of sensors across the 140 buildings at the EXPO. These systems prove the concept and demonstrate the feasibility of green and smart technologies related to integrated building and district management systems and practices at scale.

Grid edge

Solutions at the "Grid Edge" enable buildings, infrastructure, and industries to interconnect and interact to optimize energy efficiency and create new businesses. By placing computing power closer to the user, users can intelligently integrate renewables and take control of their energy supply.

Energy savings can power vehicle-to-building or building-to-grid platforms, which can identify the lowest-cost combination with the least emissions source or even offload reserves into battery storage. With grid edge technology, consumers become prosumers who produce, store, and use electrical energy. This not only facilitates the decarbonization and decentralization of energy systems but also creates possibilities for a variety of innovative business models, including off-peak rates, demand response management, and community shared renewable energy.

Water management

A key global sustainability challenge, especially in arid regions, is the supply of freshwater for human consumption and basic sanitation, as well as agriculture to ensure food security.

At EXPO2020, Siemens demonstrated a smart irrigation system that ensures less water is wasted by monitoring and controlling it via the MindSphere industrial IoT platform. In such a system, IoT devices tell irrigation systems exactly where and when to water and how much water to spray as well as the type of vegetation to grow, depending on atmospheric conditions.

A smart irrigation system allows data to be collected and analyzed intelligently so that control pumps/valves can be monitored and controlled, water usage tracked, and leakages detected in the system. Further extending this throughout the Expo site in correlation with weather data will mean such savings could also be enjoyed throughout the exposition.

Green hydrogen

Hydrogen generated via non-fossil means — referred to as "green" hydrogen (H_2) — is fast gaining attention and investment as the search for zero-carbon energy accelerates. To demonstrate this technology, EXPO2020 Dubai, Siemens, and the Dubai Electricity and Water Authority (DEWA) inaugurated the region's first green H_2 facility in May 2021.

With clean energy a key pillar in the United Arab Emirates' "Net Zero by 2050" strategic initiative, this collaboration established a 10,000 m^2 green H_2 facility at the Mohammed bin Rashid Al Maktoum Solar Park in Dubai. During the day, the plant harnesses some of the photovoltaic electricity from the MBR Solar Park to produce about 20.5 kg/h of green H2 at 1.25 MWe of peak power using a technology called "Polymer electrolyte membrane" (PEM) electrolysis. At night, green H2 is converted into electricity to power the city with sustainable energy. The facility will serve as a testbed for the production of green H2 on an industrial scale to serve as a buffer for renewable energy production, both for fast response applications as well as for long-term storage. It is hoped that this public-private partnership will also serve as a model for how the private sector and governments can collaborate to promote sustainability and innovation.

At COP26 in Glasgow, Siemens Energy joined 50 regions and more than 26 organizations in signing the statement on "Global coal to clean power transition strategies", which aims to scale up clean power and strengthen domestic and international efforts to phase out coal. It was therefore worthy that as delegates departed from the conference, the Dow Jones Sustainability Index ranked Siemens as the most sustainable company in its industry group for 2021, repeating its annual ranking in this prestigious index since 1999. Its efforts at Dubai EXPO2020 is a vivid demonstration of the commitment to sustainability it has made and how private corporations can also play a critical role to limit global warming to 2°C or less.

INTERVIEW
Dr. Cary Chan
Executive Director, Hong Kong
Green Building Council

by Eric Chong

An interview with Ir. Dr. Cary Chan, JP,[47] Executive Director of the Hong Kong Green Building Council and Chair of Asia Pacific Network of World Green Building Council.

Cary, you are a pioneer in the sustainable built movement space, having played a leading role in its expansion over the years. What inspired you to dedicate your time and effort to this cause?

I began my involvement in sustainable built environment in 1995 when I joined Swire Properties to establish HKBEAM, the world's second green building assessment scheme.

At that time, it was simply to do the job assigned to me. This, however, allowed me to work on energy saving for the company's portfolio, which I did to impress on the company the operational cost savings potential that sustainable buildings could bring.

This work sparked my interest to research in the area of knowledge-based energy management, with the aim to improve the energy efficiency of buildings. One can say that at the time I was driven by my own interest.

It was not until early 2000 that I became aware of climate change and started to consider this a social responsibility issue and contribute to meeting this global challenge. Motivated by this, I actively promoted knowledge-based energy management, sharing my knowledge and experience with everyone in the building industry. I established many initiatives to engage more stakeholders, such as offering free energy audits to our tenants, something not too common at that time. These efforts were well recognized by the industry.

This provided further impetus for me to join the Hong Kong Green Building Council (HKGBC) as their executive director in 2016 to have an impact on stakeholders beyond Hong Kong, such as in the Greater Bay Area.

I am most honored that in 2021, the World Green Building Council (WorldGBC) appointed me as the Chair of its Asia Pacific Regional Network (APN), allowing me to work closely with all national GBCs in the Asia-Pacific region to lay the foundation for a sustainable built environment in the region for generations to come.

As the largest global organization that aims to transform the built environment, what is the vision of WorldGBC for the relevant sectors?

WorldGBC's mission is to transform the building and construction sector across three strategic areas: climate action, health and well-being, and resources and circularity. As a member of the UN Global Compact, WorldGBC works with businesses, organizations, and governments to drive the ambitions of the Paris Accord and the UN SDGs (Sustainable Development Goals). Through a systems change approach, the WorldGBC network is leading the industry towards a new zero carbon, healthy, equitable, and resilient built environment.

How does WorldGBC see Asia's position in the sustainable built environment?

Asia-Pacific is at the forefront of the sustainable built environment. Sixty percent of the world's population (4.3 billion people) already live in the region, with more than 2 billion living in urban areas. In fact, the region has 16 of the world's 28 mega-cities (cities with 10 million or more inhabitants). The urban population is expected to reach 3.3 billion by 2050, adding further demand for buildings.

Set against this backdrop of huge growth, creating buildings that are low or with net-zero carbon is essential to ensure a high quality of life for people, minimize the negative impact on the environment, and maximize economic opportunities. GBCs in the Asia-Pacific region are responding to these challenges and opportunities on the ground.

Asia offers opportunities on a global scale that at the same time come with significant challenges. What support do you think companies need to address these?

The first thing that the companies need to know is the climate risk that their organization is exposed to. The Taskforce on Climate-related Financial Disclosure (TCFD) sets out a framework for companies to assess their climate risk and put it in financial terms. Such disclosures give investors a good picture of the climate risk that the company is exposed to and how the company is managing the risk. Companies will need support to adopt TCFD within their organization and make it part of their strategic framework and operational protocol.

The next is for companies to set up their target and roadmap for "Advancing Net Zero". To support companies to achieve that, a number of GBCs have already established net zero target setting mechanisms for the building industry. This will help the industry to know the amount of reduction they need to target for a building to be called "net zero".

In Hong Kong, the HKGBC is in the process of establishing one for the local building sector.

There are numerous green building projects in Asia. Are there any that you would like to highlight? Can you share some of the thinking behind these projects?

There are many excellent examples that I can quote, but there is one I would like to highlight for its contribution to the sector. The National University of Singapore has completed a Net Zero Energy Building that has been in use for about 2 years. The building is a living laboratory and was able to achieve net zero without the need to offset or procure renewable energy outside the building's boundary. It is making a great contribution to the industry via living research and development into this new and exciting area. Some of the key features include an innovative hybrid cooling system that ensures rooms are not overly cooled, an array of 1,225 solar PV panels that can supply more than 500 MWh per year, and passive house features that provide natural light but yet allow for a cooler interior without artificial cooling.

What do you see as the trends that will drive development in the sustainable built environment space?

One major observation I have is that more and more governments will commit to carbon neutrality in the coming years. This will mean that net zero targets might be set for the building industry, which may be followed by carbon trading schemes and related building regulations. All these possibilities will result in risks for companies that are not taking action.

Another trend I see is that investors are also playing an essential role as they look into environmental risk as part of their corporate social responsibilities. Consequently, more funds will flow to zero or low carbon technologies and accelerate their adoption.

From a financing perspective, the building industry on the supply side may benefit from increasing green financing that provides more favorable terms for their green projects. On the demand side, financing tools such as green mortgages will generate more interest from consumers for green buildings and fitting out.

Whilst the industry will continue to focus on carbon neutrality, this will shift from operational carbon to whole life cycle emission, putting pressure on the value chain for low to zero carbon design and products. Target setting will extend to embodied carbon to not just cover raw material in construction but also in-use and end-of-life carbon. More research will be needed to determine baselines, measurement techniques, benchmarking, and reporting.

As the world increasingly feels the ramifications of climate change, one major development has been governments and the private sector paying increasing attention to resilience and adaptation in the built environment. Countries like Singapore have already set standards for town planning based on a certain sea level rise. In the private sector, there will be growing demand by investors for disclosure of climate-related risks in financial terms as framed by the TCFD. All these will mean that the sustainable built sector will need to have more provisions for resilience and adaptation in its development roadmap.

Technology and innovation are driving solutions that are benefiting both the industry and the environment. Would you care to elaborate on some examples?

Technology and innovation are driving positive change in the industry, benefiting all stakeholders. Excellent examples include:

- On operational carbon, it is expected that the efficiency for LED lighting will double within 10 years.

- Efficiency of today's PV panels is about 22%, while prototypes are being tested with an efficiency of over 44%. This could increase the percentage of renewable energy (RE) for a building.

- Use of "Direct Current" (DC) distribution, which can allow smart DC grids that can connect with distributed power like RE and also storage. Such systems can reduce transmission and conversion loss from AC (alternating current) or DC required in such schemes while also managing supply and demand instantaneously.

- For embodied carbon, timber-structured "Wooden Skyscrapers" are developing quickly, with buildings now over 30 floors constructed by timber, a carbon-positive material.

- Low carbon concrete using captured carbon for curing is being adopted by the industry, and carbon-positive concrete is being developed.

- Carbon capture, storage, and utilization are also being developed with the captured material used to manufacture products like nanofiber, which can be used as building material.

Climate change poses an existential threat to the world. As a leader in the industry, why and how do you think firms and governments should contribute to address this?

All firms should contribute to addressing this immense threat by not just reducing carbon emissions within the areas of their business but also to influence their value chain. For instance, a developer should adopt responsible procurement policies and practices so that a green product market can become the mainstream practice for the industry. For building owners, they should engage tenants to enable their buildings to be truly carbon neutral as a whole instead of just reducing carbon emissions in their area of responsibility.

As a whole, the industry should establish and disclose its "Advancing Net Zero" roadmap so that all stakeholders can follow it. It should also quantify risk in financial terms so that all players can have a clearer picture on the cost of not taking action. Above all, it is crucial that the government starts addressing the issue of resilience and adaptation in its town planning. This will help to mitigate the cost to future generations of not taking action today.

Endnotes

1. United Nations, "UN cities — United Nations sustainable development", United Nations, 2021, https://www.un.org/sustainabledevelopment/cities/. Accessed: 15 August 2021.
2. Statistica, "Carbon dioxide emissions worldwide from 1965 to 2020 by region", Statistica, 2021, https://www.statista.com/statistics/205966/world-carbon-dioxide-emissions-by-region/. Accessed: 15 August 2021.
3. Hannah Ritchie and Max Roser, "Urbanisation", OurWorldInData.org, 2018, https://ourworldindata.org/urbanization.
4. Economic and Social Commission for Asia and the Pacific, "Urbanisation trends in the Asia and the Pacific", November 2021.
5. Rajiv Biswas, *Asian Megacities* (London: Palgrave Macmillan, 2016), pp. 52–65.
6. Chris Sneddon, Richard Howarth, and Richard Norgaard, "Sustainable development in a post-Brundtland world", *Ecological Economics* **57** (2006) 253–268.
7. Alex Opoku, "SDG2030: A sustainable built environment's role in achieving the post-2015 United Nations Sustainable Development Goals", *Association of Researchers in Construction Management* **2** (2016) 1149–1158.
8. "Sustainable construction: Future R & I requirements — analysis of current position", Construction Research and Innovation Panel Report, March 1999.
9. Stephen Siu-Yu Lau, Zhonghua Gou, and Deo Prasad, "Market readiness and policy implications for green buildings: Case study from Hong Kong", *Journal of Green Buildings* **8** (2013) 162–173.
10. Amos Darko and Albert P. C. Chan, "Critical analysis of green building research trend in construction journals", *Habitat International* **57** (2016) 53–63.
11. Charles J. Sendzimir, Jan Sendzimir, and Brad Guy, "Construction ecology and metabolism: Natural system analogues for a sustainable built environment", *Construction Management and Economics* **18** (2000) 903–916.
12. Jun Xia, Dennis Poon, and Douglas Mass, "Case study: Shanghai rower", *Council on Tall Buildings and Urban Habitat Journal*, 2010, pp. 12–18.
13. Hong Kong Environment Bureau, "Energy savings plan for Hong Kong's built environment 2015–2025+", Environment Bureau of the HK SAR Government, 2014.
14. N. Soares, J. Bastosa, L. Alexandre, *et al.*, "A review on current advances in the energy and environmental performance of buildings towards a more

sustainable environment", *Renewable and Sustainable Energy Reviews* **77** (2017) 845–886.

15. Derek Clements-Croome, ed., *Intelligent Buildings: Design, Management and Operation* (London: Thomas Telford Publishing, 2004).

16. The Royal Academy of Engineering, "Smart infrastructure: The future", The Royal Academy of Engineering, United Kingdom, January 2012.

17. P. Suresh, J. V. Daniel, V. Parthasarathy, and R. H. Aswathy, "A state of the art review on the Internet of Things (IoT) history, technology and fields of deployment", 2014 International Conference on Science Engineering and Management Research, 2014, pp. 1–8.

18. Wiliot Inc., "Wiliot platform: IoT pixels", Wiliot Inc., https://www.wiliot.com/product/iot-pixel#01.

19. Susan Decker, "Huawei's 5G patents means U.S. will pay despite Trump ban", *Bloomberg*, 2020, https://www.bloomberg.com/news/articles/2020-06-08/huawei-s-patents-on-5g-means-u-s-will-pay-despite-trump-s-ban.

20. Jennifer Wills, "5G technology: Which country will be the first to adapt?" *Investopedia*, 20 August 2021, https://www.investopedia.com/articles/markets-economy/090916/5g-technology-which-country-will-be-first-adapt.asp.

21. Jay Lee, Behrad Bagheri, and Hung-An Kao, "A cyber-physical systems architecture for Industry 4.0-based manufacturing systems", *Manufacturing Letters* **3** (2015) 18–23.

22. International Energy Agency, "GlobalABC regional roadmap for buildings and construction in Asia 2020–2050", International Energy Agency, August 2020.

23. Jarek Kurnitski, "REHVA nZEB technical definition and system boundaries for nearly zero energy buildings", *REHV Journal* **5** (2013) 22–28.

24. Sergio Altomonte, Peter Rutherford, and Robin Wilson, "Human factors in the design of sustainable built environment", *Intelligent Buildings International* **7** (2015) 224–241.

25. Architecture 2030, "Why the building sector", *Architecture 2030*, https://architecture2030.org/. Accessed: 15 August 2021.

26. P. K. Mehta, "Natual pozzolans: Supplementary cementing materials in concrete", *CANMET Special Publications* **86** (1987) 1–33.

27. Bhupinder Singh, *Waste and Supplementary Cementitious Materials in Concrete: Rice Husk Ash* (United Kingdom: Woodhead Publishing, 2018), pp. 417–460.

28. Mineral Products Association/UK Concrete, "Our roadmap to beyond Net Zero", 2020.

29. Gammon Construction Ltd., "Sustainability report 2020", 2020.
30. Oscar Holland, "Wooden skyscraper revolution", *CNN*, https://edition.cnn.com/style/article/woodenyscraper-revolution-timber/index.html. Accessed: 15 August 2021.
31. Channel News Asia, "New S$180m building at NTU to be Asia's largest wooden building", *Channel News Asia*, 27 August 2018.
32. Gregory M. Wilson, Mowafak Al-Jassim, Wyatt K. Metzger, *et al.*, "The 2020 photovoltaic technologies roadmap", *Journal of Physics D: Applied Physics* **53** (2020) 1–47.
33. Philip Kwang, "Forum: Use vertical facades of buildings for solar power to beat land constraint", *The Straits Times*, 28 October 2021.
34. Tran Hung, Daisuke Uchihama, Shiro Ochi, and Yoshifumi Yasuoka, "Assessment with satellite data of the urban heat island effects in Asian mega cities", *International Journal of Applied Earth Observation and Geoinformation* **8** (2006) 34–48.
35. Gabriel Happle, Erick Wilhelm, Jimeno Fonseca, and Arno Schlueter, "Determining air-conditioning usage patterns in Singapore from distributed, portable sensors", *Energy Procedia* **122** (2017) 313–318.
36. Passive House Institute, "What is a passive house", Passive House Institute, https://passivehouse.com/02_informations/01_whatisapassivehouse/01_whatisapassivehouse.htm. Accessed: 15 August 2021.
37. Anne Jacobs, "How passive house was adopted in Asia", *Commerce and Communications to the Point*, Issue 1, 2017.
38. U.S. Green Building Council, "LEED zero program guide", April 2020.
39. Dannielle Torrent Tucker, "In a warming world, Cape Town's 'Day Zero' drought won't be an anomaly, Stanford researcher says", *Stanford News*, November 2020, https://news.stanford.edu/2020/11/09/cape-towns-day-zero-drought-sign-things-come/.
40. Ezzat A. A. Othman, "Corporate social responsibility of architectural design firms towards a sustainable built environment in South Africa", *Architectural Engineering and Design Management* **5** (2009) 36–45.
41. Eric Chong, "Influences on diffusion of eco-innovation: A study on green-building adoption in Hong Kong", DBA thesis, City University of Hong Kong, 2019.
42. Robert Simons, Eugene Choi, and D.M. Simons, "The effects of state and city green policies on the market penetration of green commercial buildings", *The Journal of Sustainable Real Estate* **1** (2009) 139–165.

43. Taryn Fransen, "Making sense of countries Paris agreement climate pledges", World Resources Institute, Washington D.C., 2021, https://www.wri.org/insights/understanding-ndcs-paris-agreement-climate-pledges. Accessed: 1 January 2022.
44. Siemens A.G., "Sustainability report 2021", 2020.
45. Siemens A.G. , "Expo 2020 Dubai: A sustainable site", 2020.
46. Office of Electricity, "Demand response", Department of Energy, Washington D.C., https://www.energy.gov/oe/activities/technology-development/grid-modernization-and-smart-grid/demand-response. Accessed: 2022.
47. Justice of Peace.

Chapter 6
TRANSPORTATION

by James Kruger, Davide A. Nicolini
and Tony A. Verb

ANALYSIS

by James Kruger and Davide A. Nicolini

Greenhouse gas emissions from transportation are second only to power generation.

But there is now significant cause for optimism. This chapter explains why. We focus mostly on road transport, for which a meaningful, global decarbonization trend is now underway. It is not that we want to dodge the discussion on aviation or shipping or any other mode of transport. It is simply that road transport is the most significant aspect of transport's overall greenhouse gas (GHG) emissions, and the decarbonization pathways now underway will shape and inform other transport pathways.

Environmentally, all modes of transport have an impact beyond carbon dioxide. Hydrocarbons and nitrogen dioxides combine in sunlight to form a layer of ozone, which is protective when high in the atmosphere, but damaging at street level. Our Asian cities' smog and respiratory health problems, especially in children, is a consequence of this. According to the World Health Organization (WHO), air pollution kills an estimated 7 million people worldwide every year. The global cost of these transportation-attributable health impacts in 2015 was approximately USD 1 trillion.[1]

Just to add to the macabre task of reporting life impacts from transport, the WHO counts 1.3 million people annually as having their lives cut short by road traffic accidents.[2] The future of transport, with autonomous driving technology, seeks to prevent all these tragedies as well. So, as we shall see, road transport is a significant enabler of a Carbonless Asia future.

PAST

One hundred years ago, people did not want faster horses — "they wanted less horseshit".[3] The change was dramatic, more so than a street merchant in New York City in 1910 could ever imagine.

Henry Ford did not simply substitute cars for horses. His cars created entirely new trade avenues and products that served previously unmet needs. Indeed, it is arguable that the very first heartbeat of modern urbanization started with the first turn of the Ford Model T starter crank.

Transport and infrastructure have always been inextricably linked. We have seen this in our lifetimes with Asia's rapid urbanization. We see it now with Supercharger networks. Significant public-private infrastructure programs built roads, bridges, airports, and rail and ferry lines and mobilized significant parts of the Asian population out of isolated rural settings and, indeed, out of poverty.

The growing smog on roads did trigger some change. Incremental improvements in engine efficiencies and a lighter weight chassis design led to significantly better fuel economies in cars and motorbikes, as well as cleaner diesel technologies for trucks and lorries. Over the last 40 years, particulate emissions for gasoline engines decreased by a factor of 100.[4] However, the consumer desire for SUVs — which the Asian middle class also followed — largely offset these efficiency gains.[5]

And then there was "Dieselgate".[6]

PRESENT

In China, rapid urbanization was causing massive GDP-sapping city congestion, blanketing pollution, and a prevalence of respiratory health issues. The central planning government, with their seemingly limitless new infrastructure budgets and political will, coordinated the build-out of a new type of road traffic and, in that process, the development of entirely new and globally significant industrial companies. Chemical processors, magnet sintering plants, battery fabricators, electric vehicle makers, and charging network operators all fed into a burgeoning local ecosystem. Chinese battery maker, CATL, is now worth more than General Motors and Ford combined.

While the West was concerning itself with Dieselgate and/or dismissing Elon Musk's Tesla, China was moving the playing field.

Municipal bike-sharing, once just a Paris phenomenon, gained popularity in China. Innovative private bike-sharing companies — such as Ofo and Mobike — expanded very rapidly at the start of the Electric Vehicle (EV) wave. The first generation of their micro-mobility offering was riddled with problems for both the company and China's city planners: wasted bikes and scooters piled up in massive metal "graveyards". However, the technology stack was quickly improved: Internet of Things (IoT) enabled asset tracking, ID verification, smart locks, geo-fencing, mapping and navigation, real-time fleet data and analytics, and intelligent fleet management and charging. All these were augmented with a better user experience via better payment apps and integrations with mass transit networks.

Throughout COVID-19 and ahead of COP26, many western countries started to recognize the economic and environmental template offered by China for an orderly transition away from the polluting combustion engine. However, the geopolitics on the global scale was less straightforward, and there was no top-down central planning government, with the notable exception of Norway, ambitious enough to overcome all the headwinds and distractions. While the South Korean and Japanese chemical engineering firms (Posco, LG, SK, Panasonic, and so forth) were busy setting up factories in China, their expansion into Europe and America was slow. South Korean LG sued SK[7] in 2019 for IP infringement in what became a multi-billion, multi-year distraction for the decarbonization of transport beyond China.

Further, the Fukushima-inspired denuclearization programs in Japan, Germany, and Korea, to a lesser, slower extent, increased the risk of baseload electricity grids and made them reluctant to invest in EV-charging infrastructure. It was an unhappy coincidence that these countries happen to be where most carmakers reside.[8]

In 2021 things changed. Most of the world's biggest carmakers declared the end of combustion engines. And SK and LG settled their litigation, which enabled them to establish separate entities for their battery businesses and embark on significant capital investment in the supply chain. The publicly listed carmakers became forthright and absolute in their pivot to battery EVs; their stakeholders and institutional capital support no doubt prodding them along the way, with their increased focus on publishable Environmental, Social, and Governance (ESG) metrics and the performance of Tesla's stock price. Those companies with less public market pressure for ESG results were still appearing to hedge their bets across a range of decarbonizing technologies, hydrogen fuel cells included. Toyota and BMW come to mind: Toyota is run today by the founder's grandson; BMW has 50% of its shareholding privately held by two wealthy families.

Electric mopeds, whether as part of a mobility-as-a-service offering or privately owned, started becoming popular in Asia. As we discuss later, battery swapping is a significant ease on the infrastructure build-out and a tailwind for electric moped adoption. This is great news for the environment. According to a Wired magazine comparison,[9] one kilowatt-hour of energy can only get a gasoline-powered car to travel 0.8 miles versus 4.1 miles for an EV, compared to 82.8 miles for an electric moped.[10]

More broadly, transport and, in particular, how individuals engaged their transport needs, started to be looked at differently. Ride-hailing services emerged and gained strong support from the capital markets because of their disruption and growth narrative. Grab and Gojek in Southeast Asia used the act of hailing rides amongst dense urban populations as a platform for dominating the attention of eyeballs on mobile phone screens. They added food delivery, digital wallets, payments, and other financial services into their "Super Apps". This improved their scale and unit economics beyond their western counterparts, in particular, Uber and Lyft, who have always needed autonomous technologies for robo-taxi fleets as their end game.

Micro-mobility options soon appeared across all Asian cities to assist with city traffic congestion. To illustrate with India, the startup Bounce now operates more than 20,000 electric and gasoline dockless bikes and motor scooters and is anticipating significant growth post-COVID-19. Today, shared bikes are now the third-most popular mode of public transit in China.

To fill some of the voids of rushing into car ownership as a young adult, app-enabled peer-to-peer car sharing have emerged, such as Socar in Korea, ASSA Mobility in Indonesia, and Car Next Door in Australia.

Overall, there is a global frenzy in mobility and transport, thanks to technology, government programs, public desire, and increased spending power on mobility-as-a-service "fueled" by a growing middle class and a trend of delayed car ownership.

FUTURE

One day — and probably sooner than we think — we will look back with bemusement on our lived experience with cars. How strange those days will seem when we took out high-interest loans to buy cars, and fill them up at service stations with retail margins on the fuel, and then drive them into the city during peak hours, and let them idle with noxious exhaust fumes blowing during traffic congestion. Perhaps even stranger still, we will look back at how we would park our vehicles at expensive, privately run car park buildings and pay excessively for the privilege of having the car sit there (and remain unutilized for 95% of the day), all the while it would be depreciating rapidly in capital value.

Clearly, our lived experience with transport is highly disruptable.

As many of Asia's megacities follow Singapore and London's lead with congestion tax systems, driving to work in one's own car will definitely become more and more the exception rather than the norm. Through this, micro-mobility options will emerge as ubiquitous. By themselves they are a convenient, affordable, and relatively safe mode of city travel. And soon, they will become completely embedded within the city infrastructure, with broad public-private partnerships and bundled ticketing within the various public transit systems.

SoCar in South Korea, for example, will be aiming to integrate all the mobility options for the Korean city traveler into a "super app", providing users the best point-in-time option across all modes of transit "anywhere, anytime". This includes car-sharing, buses, trains, and electric bikes.

Also changing the cityscape will be autonomous delivery robots for food and last-mile e-commerce deliveries. Meituan, China's biggest food delivery company, is now trialing unmanned robots in offices and hotels in Beijing and Shenzhen. In Japan, the robot company ZMP has developed and is now testing a food delivery robot in universities and apartment complexes. Indeed, all the major e-commerce providers, like JD.com in China, have programs to expand autonomous mini-vans for their last-mile delivery requirements.[11]

We may also see small electric helicopters or e-VTOL crafts[12] becoming a commute or short-haul transport option for certain travelers. China's Ehang, headquartered in Guangzhou, is building an entire urban air mobility ecosystem. In a way, it could be said that the three big mobility tech themes of electricity, connectivity, and autonomy are actually the three "horsemen" of the privately owned petrol car's apocalypse.

McKinsey has estimated that 50% of the transport industry will be disrupted by 2030.[13] If that holds, one in two transport companies running today will disappear in under ten years. Their only option to survive will be to evolve out of fossil fuels and embrace and not underestimate the sheer enormity of the disruption. It will be as significant a disruption as the one Henry Ford inflicted on the horse and carriage industry and one that will move beyond cars, trucks, and road transport.

Looking to the Skies or Out to Sea

Shipping and aviation collectively contribute to about 5% of all GHGs worldwide. Even though there is hope that external factors will cause a drop in their demand,[14] technology will be required to effect decarbonization. But they are more complicated to decarbonize. Batteries cannot do the heavy lifting; lithium-ion batteries do not have the energy density per weight.

One needs roughly 3 GWh of energy to fly a jumbo jet from London to Singapore. With current battery technology, that equates to 10,000 metric tons worth of battery mass. To put it another way, for batteries to replace a current jumbo jet's fuel mass, which equates to 200 metric tons, the specific energy improvement in batteries from where we are today would need to be 50x. One cannot reasonably imagine batteries to solve this for the airline industry in the same way they are currently doing for cars.[15]

The rate-limiting factors for decarbonizing technologies in aviation are the energy density of the fuel load and, in the case of passenger aircraft, where aerodynamic drag is such a critical economic factor to the profitability of a flight, the overall volume of that load. Further, as planes and ships travel ubiquitously from Port A to Port B (and may need safety divergence to other adjacent ports), they need the same fueling systems and easy-to-build, ubiquitous infrastructure.

There will possibly emerge a dichotomy in the markets of short haul (say 500 miles) versus long haul (cross ocean, Japan to Singapore, etc.). Firstly, short-haul aviation and sea transit will be competing somewhat, with cheap land-based shared transport linked by new bridges and highways.[16] Secondly, there is no obviously superior propulsion technology for long-haul transport, especially when weighed up against the relatively cheap, energy-dense hydrocarbon fuel. And we all know unit economics matter. This might change, perhaps by 2040 or by the time material carbon taxation becomes a global reality.

Because of these elements, technology disruption has not yet emerged.[17] One interim measure is bio-fuels. They will not be needed in road transport following its electrification, so we can soon cater for the redirection of the current supply of bio-fuels[18] to aviation and shipping. Other forms of synthetic aviation fuels[19] could supplement this supply, but they are currently very expensive. We will need economies of scale or carbon taxes to make them viable.

When? Forecasting Needs Systems Thinking

In 2018, we had the benefit of spending significant time with the C-Suite of Didi, one of the largest ride-hailing operators in the world, and Geely Auto, the largest privately owned Chinese carmaker. They were, and still are, both operators of sizable car fleets in the country; Geely owns a premium limousine hailing service called "CaoCao". It was obvious to these executives *back then* that the future of their car fleets was electric. Their assets would be easier to finance to start with. The lithium-ion battery pack was an expensive component then and now — some 25% of the overall cost — but it was modular and extractable for a valuable second life as energy storage for 5G telecom towers. EV fleets would also prove easier to maintain with the far fewer moving parts.

Furthermore, technologies around data and artificial intelligence (AI) would enable their future plans for robo-taxis. Semi-autonomous and autonomous driving and data-connected cars would be better enabled by an EV architecture — the shell design, sensor placement, and chip deployment could be redrawn from scratch. There are already robo-taxi companies in live test operations in China, such as Baidu's Apollo Go and Alibaba-backed AutoX in Beijing and Shenzhen.[20]

These same Didi and Geely executives could, in 2018, also see where the electricity grid was heading. The technological advancements in the "renewables electricity stack", such as solar, wind, grid management, storage, rooftop solar, and home energy storage, and the technologies in the "propulsion stack", such as batteries, EV drive chains, and others, would enable each other. A decentralized, digitalized electric energy grid — a source of new and second life battery demand — would help the adoption of electric cars and vice versa.

Around the same time, one of the authors was also an attendee of many Oil & Gas industry strategy presentations. The differences of opinions between these forward-looking car companies and the petrol companies were stark. As we can see from Figure 6.1, taken from an investor presentation, oil executives were viewing oil demand from transport as something they could bank on well into the future, a commodity that would continue to grow out at least until 2040 and perhaps beyond. Maybe, they thought, a 50% sale of cars by 2040 *might* be electric, but that there would still be plenty of flammable juice left for their segment of the market.

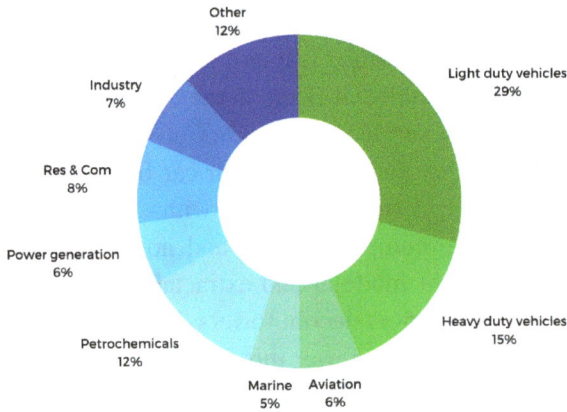

Figure 6.1.   2040 Petroleum Industry Forecast from 2018: Oil still dominating transport and petrochemicals.

The mistakes by the Oil & Gas forecasters are now obvious as we can see from the uptake in EVs in Figure 6.2. Just as the Horse & Carriage industry in the early 1900s underestimated its new technology threat, it is happening again within the petroleum industry. It seems that it is common for human nature to narrow perspectives on future risks. With technology

Global annual sales of plug-in electric passenger cars in top selling markets (2011-2021)

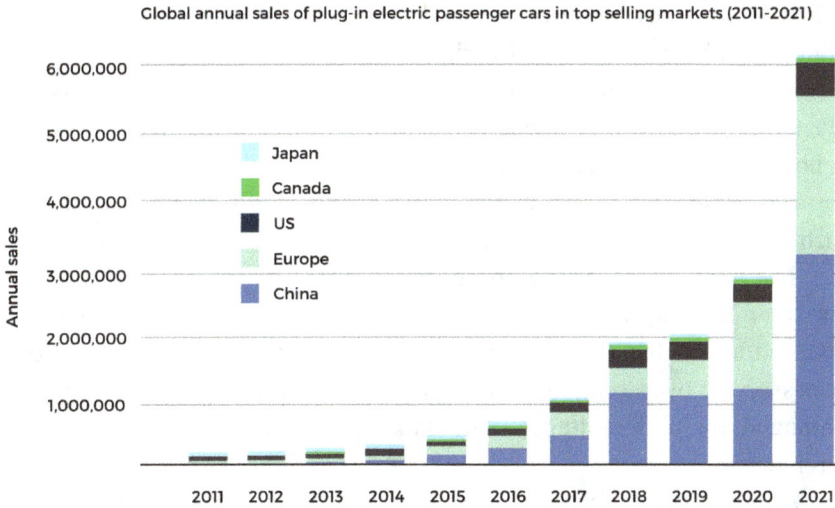

Figure 6.2. Electric vehicle sales by country.

disruptions, these cognitive biases lead to an underplaying of the learning rate for the incoming technology, underestimating the pioneers' product evolution from the first to subsequent generations, and failing to see the ecosystem converging around the new technology. These ecosystems provide the economic tailwinds, establish enabling infrastructure, and accelerate the learning rates.

This is not to pronounce lithium-ion as the decarbonizing panacea for all transport. Just as it has been underestimated, it should not now be overestimated. One cannot deny the laws of gravity. Let us therefore look at all the propulsion technologies that will assist to decarbonize transport.

Batteries

The breeding ground for lithium-ion batteries as the key component of an electric propulsion technology in cars was years of development in consumer electronics, power tools, and mobile phones. The likes of Panasonic, Samsung, LG, and SK were household names developing technologies for our pockets. The crucial enabling feature for lithium-ion batteries, unlike lead acid batteries, was that they could be broken down into a single modular cell. They were usable in different ways and with different sizing. And so "lithium-ion" became a broad design category of technologies with a vast array of different materials; in the cathode, anode, separator, and binding agents and in packing techniques. Lithium-ion is not a single entity, and there is no one "killer app" within the lithium-ion category. In reality, the different formats and chemistries are compromises and trade-offs in one performance metric versus another, whose trade-offs are driven by the application.

The modularity and flexibility of lithium-ion batteries were critical for their learning rate across their transport applications. Engineering adaptations to the cell design and battery management system were necessary to overcome the tendency for lithium-ion to overheat and be damaged by high voltages. The inherent modularity facilitated this engineering.

Another key enabling feature is lithium's relative abundance in the Earth's crust. It is true that for any energy storage and propulsion revolution to be credible today at a global level, it needs to be scalable. And we should also add: "sensible" from a "well-to-wheels" climate impact perspective.

Despite projected higher raw material input prices, there are cost reductions and performance improvements to come. Figure 6.3 shows

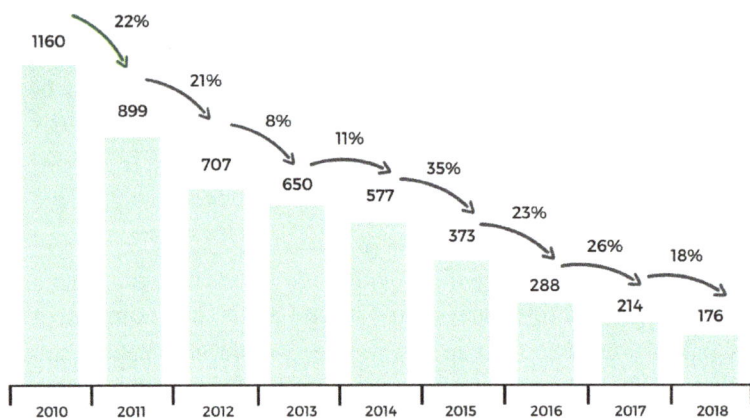

Figure 6.3.   Plummeting costs of batteries.[21]

Bloomberg is projecting the cost of a lithium-ion EV battery pack to be below USD 100 per kilowatt-hour by 2023.[22] This price is often regarded as the benchmark when the sticker price of Electric Cars is cheaper than combustion cars.

CATL and BYD have given China a commanding position in the battery supply chain. These companies already supply batteries to almost all of the world's automakers (including GM, Volkswagen, BMW, and Tesla) and have been very strategic in securing vast supplies of the raw materials that go inside the batteries. That supply chain dominance has now stirred fears in Washington about local supply lines and production capabilities. Rightfully so, China has approximately 14 times the electric car battery-making capacity of the United States.[23]

BYD not only boasts of having Warren Buffet as a significant shareholder, but they possess an impressive portfolio of adjacent manufacturing and technology capabilities. Just like the development of the technology itself, BYD started as a consumer electronics battery manufacturer. It then quickly broadened, moving into solar panels production, large-scale battery projects, and electrified cars, buses, and trucks.

Its market-leading position in electric buses deserves particular attention. They have global prominence and will continue to gain global market share for the foreseeable future.[24] According to Bloomberg, there were about 598,000 e-buses on roads globally at the end of 2020, of which 585,000 were in China (primarily shared between BYD and its Chinese rival Yutong).[25] BYD is now accelerating production and receiving many orders from municipalities from all parts of the globe — from Finland to Israel to Columbia.[26]

In the next decade, buses should prove to be mostly electrified. Bloomberg predicts that the number of e-buses on roads will reach nearly 1.7 million in 2030. Arguably, this is a conservative estimate. Since buses have predictable routes and significant rest periods (enabling slower charging), the infrastructure built for electric buses should prove to be incremental. Further, buses enjoy government financing and other concessions, so the total cost of ownership over their lifetime (including maintenance and wear and tear, where EVs have an advantage given their fewer moving parts) is more important than just the "sticker price" upon acquisition.

Hydrogen Fuel Cells

Turning our attention to hydrogen, its molecular characteristics provoke excitement in many engineers:

- Hydrogen is energy-dense (3x gasoline and gas) and light (one-tenth the weight of natural gas). However, it is only 25% of the energy per unit volume of natural gas, whether liquefied or as a gas. This is important and the reason why hydrogen is stored at high pressure (up to 700 times atmospheric pressure).

- It can be produced anywhere with electricity and water: Nine liters of water with 50 KWh of electricity can produce 1 kg of hydrogen, which would propel a car for about 100 km.

- Hydrogen can generate either heat or electricity to deliver power. On paper, it can be 100% efficient when burned, unlike fossil fuel, albeit it delivers only 60% efficiency via a fuel cell.

- Hydrogen can be produced, stored, transported, and used without $CO2$ emissions. However, because of its extreme lightness and lack of energy density by volume in its natural state, storage and transport require further, energy-intensive processing.

- It burns at a similar temperature to natural gas.

Arguably, since hydrogen fuel cells can extend a vehicle's range to over 1,000 km, their practical limitations may be "worth it" in certain cases. From a desktop technical comparison, hydrogen fuel cells can store a lot more energy onboard. Accordingly, they have more range than comparable-sized battery EVs.[27]

A fuel cell may prove viable from the perspective of a weight-to-power ratio, cost, and other factors. This is why one should not speculate on an EV-only future. For certain applications and in certain localized, closed-loop infrastructure build-outs, it may come down to the energy needed onboard the vehicle, where the power to weight will matter more. In Australia, companies like Fortesque are speculating on a fulsome future demand for green hydrogen. One obvious use in transport for them is heavy haulage by rail and truck from a remote part of the Western

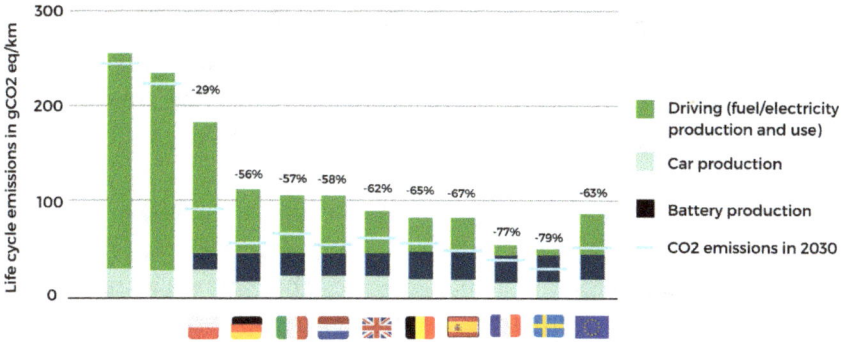

Figure 6.4. Today's petrol and diesel cars emit almost three times more CO2 than the average EU electric car.[28]

Australian Pilbara to the relevant ports (and then potentially on the container ships).[29] We have no cause to doubt their ambition. On the other side of the Australian continent, Korea Zinc is ordering Hydrogen Trucks for its Queensland smelter.

Typically, for a government to support a transport initiative, the defining parameters will ultimately come down to the cost of the infrastructure build-out and the extent to which that cost can be shared across many applications with ubiquitous adoption. For this reason, it seems most likely that lithium-ion-battery-powered cars and buses (and we would argue trucks) will be the vast majority of vehicles on Asian roads. As we see from Figure 6.4, even with the sub-optimal grids (i.e., not maximized for renewable generation) that we currently have, this is good news for the environment.

Alternative Fuels

The most well-known alternative fuels are biofuels, which can be made from many different types of feedstock with varying degrees of compromise to the environment and varying hopes for technological learning rates:

- oils, fats, and industrial wastes,
- crop-based feedstock as a bioethanol or biodiesel, and
- new technologies using microalgae.

Today's advanced biofuels are a lot different from the first-generation versions, such as ethanol. Some are made from plants that are not grown for food, so they need little to no fertilizer. Others are made from agricultural by-products, like corn stalks and the pulp left over from making paper.

Another type of alternative fuel is electrofuels (or e-fuels) — which is not a biofuel — and it uses electricity to combine the hydrogen molecules in water with the carbon in carbon dioxide. The carbon dioxide of this process is captured directly from the atmosphere, so burning electrofuels does not add to overall emissions. However, they are very expensive and can cost anywhere from 3 to 7 times as much as fossil fuels[30] and so are unlikely without significant carbon taxing and central incentivization. But this is a technology to consider for the future if economies of scale make it competitive.[31]

While biofuels are unlikely to be a permanent solution for any particular transport mode — and many environmentally minded futurists would protest their promotion and increased use — they will likely have a significant place as a bridging fuel. Worldwide biofuel production is currently in excess of 160 billion liters per year, and if one can imagine road transport being electrified or hydrogen fuel-celled, a significant amount of biofuels could be liberated as a "drop-in fuel source"[32] for shipping or long-distance aviation. One clear advantage of any propulsion technology option that centers around a "drop-in fuel"[33] is that there is no reconfiguration of the fuselage or massive capital expenditure and safety clearances from craft redesigns.

It is possible that the supply of ethically and sustainably produced biofuels are not sizable enough to warrant sufficient consideration. For that reason, there is a fair degree of alarm around them. But it is an appealing proposition to explore, working appropriately with the industry and ensuring a greater level of transparency on net environmental impact. India's agriculture waste, for example, with an estimated 900 million tons, would be enough for conversion to a new supply of 300 million tons of biofuel — more than doubling the current production. Many Asian countries have tons of grain and nut waste that could be used to create a very clean biofuel. Further, any developments in e-fuel efficiency or biofuels from microalgae would have a large addressable market for the medium term.

Innovation in Road Transport Feeds Industrial Complexes

Many industries feed into the modern car. It is a vast array of chips and circuitry, form factors, components and materials science, production lines, interwoven supply chains, and, most recently, software and AI for self-driving cars. It is about as complex a mass-consumed global product there is. Looking at the global rankings of countries' localized economic complexity represented in Figure 6.5,[34] one can quickly gauge where future car industries will be located — where the local economy is sufficiently diverse and has enough embedded capacity to deal with "complexity" and productive knowledge.

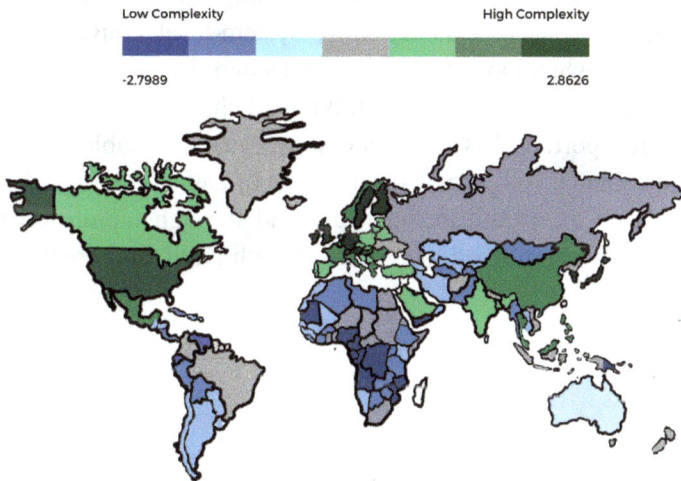

Figure 6.5.   Global map of economic complexity by country.[35]

Local country leaders across Asia would be well advised to continue efforts to support a diverse technology base in their backyards. It is going to be an increasingly relevant factor in economic growth.

CONCLUSION

Transport is important for decarbonization. But working out the decarbonizing pathways is tricky. It is tempting to immediately extrapolate out the electric car trajectory in one country for other's road transport — such as long haul trucks — and then, more broadly, for all other forms like trains, ferries, ships, planes, and the rest. It is tempting to view the foundation of electric cars as an immediate basis to electrify everything in transport.

But, different modes of transport present different technology applications, infrastructure needs, and economic adjacencies. We predict that long-haul aviation and shipping are unlikely to be solved by battery electric technologies — the decision about which "energy vector" to use is enormous and often highly contentious. One thing is clear: the link between transport and infrastructure. Proponents of both electricity and hydrogen recognize that the choice of energy systems for transport interacts with the overall energy economy, including electric power, all modes of transport, and heat. Consequently, system-level considerations are always needed when looking at transport technologies.

Asia will be at the forefront in solving all challenges involving decarbonizing transport. It has the technology, smarts, established industry players like China, Korea, and Japan, pools of capital, motivated governments, as well as significant local demand through rapidly urbanizing populations and established megacities in both North and South Asia.

STARTUP CASE STUDY
Gogoro

by James Kruger and Davide A. Nicolini

Gogoro is the largest scooter company in Taiwan but is best known for its unique swappable battery technology that lets users avoid the usual wait for a charge. The system allows users to buy scooters without buying the battery, which is "rented" and swapped via an annual subscription charge. The company states that over 147 million Smart Batteries have been swapped since 2015, making the Gogoro Network the biggest battery swapping platform in the world. The Gogoro Network of battery swapping "GoStations" has grown to over 2,000 locations in Taiwan alone.

Battery swapping has had a chequered past. In 2011, Better Place attempted to redefine driving in Israel with a swapping platform. Even though it had an agreement with Renault-Nissan to manufacture 100,000 electric cars tailored to Better Place specifications, it only sold about 750 within 2 years while piling up losses of more than USD 500 million. It filed for bankruptcy in 2013.

What is different with Gogoro, apart from sheer market timing — Better Place was just too early in the electrification movement and use of mobile-based subscription models — is the application of batteries for mopeds or "electric scooters". The batteries are much lighter and can be handled individually by the customer like a bag of groceries. Swapping can be effected in a matter of seconds. The footprint to deploy a rack of batteries is smaller; therefore, the cost to rapidly expand swap locations is significantly lower. The power requirements to charge the battery are much lower.

Also, the user experience of an electric scooter is a significant improvement on their combustion engine counterpart — no tailpipe emissions to breathe in while waiting at traffic lights and no hot engine under the seat, a big factor in temperate Asia. App-based ignition via fingerprint or PIN code to start makes the scooters pretty theft-proof. And electric propulsion has an energy efficiency, which means they are faster and more "zippy" off the mark than most combustion scooters. And then there is the sound of silence — the sheer peace and harmony of a quiet ride.

Temasek, Sumitomo, and Engie appear to have backed a winner with Gogoro. Hot on the heels of announcing significant deals in Indonesia (with Gojek), India (Hero MotorCorp), and China (Yadea and DCJ), it announced in late 2021 an SAPC deal for a US listing sometime in 2022 with an implied value in excess of USD 2 billion. Gogoro will not be without fierce competition, however. Many players are keen to penetrate Indonesia, a market with over 110 million motorbike owners, and other Southeast Asian countries. ION Mobius, for example, will be a strong competitor and will seek to differentiate themselves on performance and quality. Their electric mopeds have a top speed of up to 110 km/h and accelerate from 0 to 50 km/h in under four seconds.

Figure 6.6.   A Gogoro battery swapping station in Taiwan, as published on https://www. gogoro.com.

CORPORATE CASE STUDY
Geely

by James Kruger and Davide A. Nicolini

Headquartered in Hangzhou, China, Geely today has owned and managed a number of brands: Geely Auto, Lynk & Co, ZEEKR, Geometry, Volvo Cars, Polestar, Lotus, London Electric Vehicle Company, and Farizon Auto. In 2020, it sold over 2.1 million cars produced from manufacturing plants in China, the United States, the United Kingdom, Sweden, Belgium, Belarus, and Malaysia. Its ride-hailing service, CaoCao, operates an electric vehicle fleet with 2 million rides ordered daily in 57 Chinese cities.

Geely Holding has also been developing cutting-edge technologies in new energy, shared mobility, vehicle networks, autonomous driving, onboard chips, low-orbit satellites, and laser communication to embrace the upcoming multi-dimensional mobility ecology. It employs over 20,000 researchers in five tech hubs across the globe (see Figure 6.8).

Geely has a long track record of identifying and executing strategic acquisitions, all the while respecting local geopolitics and allowing management the freedom to operate. The company encourages management to look for ways to make the commoditized hardware build (in the case of a passenger car, the vehicle chassis) more cost-efficient. The Volvo acquisition in 2010 and the major shareholder acquisition of Daimler in 2018 are

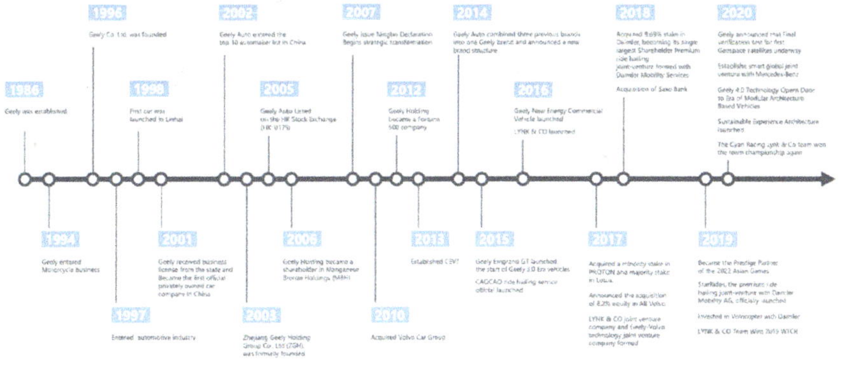

Figure 6.7. Geely's key milestones. Reproduced with permission from Gly Capital Ltd, a subsidiary of Geely Group.

| Hangzhou, China | Ningbo, China | Gothenburg, Sweden | Coventry, UK | Frankfurt, Germany |

Figure 6.8. Geely's global research and development (R&D) centers. Reproduced with permission from Gly Capital Ltd, a subsidiary of Geely Group.

cases in point. Initially, these acquisitions received some localized and politicized backlash, but Geely was patient and respectful. Now, the relationship with Daimler expands into joint ventures around the Smart car brand, heavy vehicle technologies, and limousine services. Volvo recently relisted in its native Stockholm.

Some other notable forays into the decarbonization technology strategy include:

- In February 2016, it formally incorporated a "new energy and clean alternative fuel commercial vehicle development company" pursuing four technological pathways (battery, hybrid, alternative fuels, hydrogen) for heavy vehicle transport.

- In 2017, Geely began developing a battery-swapping ecosystem, and today, Geely battery swap stations have been put into operation in several Chinese cities.

- In 2019, Geely led the Series C round for the German urban air mobility pioneer, with Volocopter raising a total of EUR 50 million (Daimler also invested; both Geely and Daimler each held a 10% stake). Later that year, Volocopter successfully completed its first public manned flight in Marina Bay, Singapore.

Geely is a truly global company leveraging strong manufacturing economies of scale within China, understanding the value of localized brand ownership and development, and investing heavily in a vast array of decarbonization technologies. With Polestar and Volvo having relisted in 2021, the company will continue to deploy capital to innovate and decarbonize city transport.

INTERVIEW
Dr. Jacob Kam
Chief Executive Officer,
Hong Kong Mass Transit Railway

by Tony Á. Verb

As a globally recognized innovator in public transport and railway operations the actions of the Hong Kong Mass Transit Railway (MTR) are great signals for where the industry is heading. We had the pleasure to conduct an insightful interview with Dr. Jacob Kam, MTR's Chief Executive Officer since 2019, to better understand the vision and mission of the company.

Whilst the MTR is a household name in Hong Kong and a significant rail and metro operator around the world, some of our readers might not have heard of it or used its services. Dr. Kam, please introduce the company as an opening.

The MTR is a world-class operator of sustainable rail transport services headquartered in Hong Kong and listed on the Hong Kong Stock Exchange. The Corporation mainly focuses on the investment in and construction and operation of mass transit passenger railways in Hong Kong, the Mainland of China, Macao, Australia, Sweden, and the United Kingdom. In Hong Kong and the Mainland of China, our business portfolio also includes station commercial businesses, the development and sale of residential and commercial properties in partnership with property developers, and the provision of property management services. Under our "Rail plus Community" business model, we not only build and operate new railway lines to enhance connectivity, but also build attractive communities along those lines, supporting the sustainable development of cities.

The MTR is one of the largest, if not the largest carbon emitter in Hong Kong. Considering that the Hong Kong SAR Government is a major shareholder in the MTR with a 2050 Net Zero target, how would you work towards this goal, and how do you work with your board?

While we do have considerable Scope 2 emissions from our purchased electricity, let's not forget that the MTR is the most environmentally friendly mode of public transport in Hong Kong. With an 8-car train being able to carry 2,500 passengers, it can carry the equivalent of 17 double-decker buses or 500 private cars. It is for this reason that the Hong Kong Government's Climate Action Plan 2050 confirms that railway will continue to be the backbone of public transport in Hong Kong.

However, we recognize the climate imperative for all businesses and, to this end, have recently completed a carbon reduction study taking into account a comprehensive range of factors, including the latest climate science, technology trends, the climate-related risks and opportunities for our business, and the views of key internal and external stakeholders.

Following this study, we have the support of our Board to set science-based reduction targets for 2030, with a longer-term goal of achieving carbon neutrality by 2050. These targets will cover the Scope 1 and 2 emissions from our Hong Kong operations as well as Scope 3 emissions and will be achieved by implementing a range of energy-saving and carbon-reduction initiatives, such as investing in the latest technologies, adopting innovative ways of working, and partnering with key stakeholders, including the electricity suppliers in Hong Kong, in a concerted effort to reduce carbon emissions. The Environmental & Social Responsibility Committee of our Board will oversee our progress towards reaching these targets.

Your company takes decarbonization seriously beyond Hong Kong as well. Please tell us about your overall position and strategy, if you have dedicated targets, and whether there is any difference on how you approach the topic in different markets?

The MTR offers a low-carbon solution that connects the communities we serve. As the operator of reliable, efficient, and environmentally friendly transportation systems, we are committed to managing our environmental and carbon footprint and to helping Hong Kong and the other cities in which we operate the transition towards becoming carbon neutral.

Our overall ESG strategy has three key focus areas — social inclusion, advancement and opportunities, and GHG emission reductions — and so we will be working with our businesses outside of Hong Kong to see how they can play their parts.

Regarding GHG emissions reduction, we have committed to setting science-based reduction targets for our Hong Kong railway and property businesses. These targets will cover the Scope 1 and 2 emissions from our Hong Kong operations as well as Scope 3 emissions from our value chain, including our franchise operations in other cities. As such, we will continue to implement carbon reduction measures in our operations in different markets. In fact, MTR Nordic has committed to setting a science-based target. It is the first Nordic traffic operator that committed to setting such a target.

Tell us about the MTR's business/carbon footprint in Asia-Pacific and globally! What is the overall consideration around that topic?

Let me just say that we present our carbon footprint in our sustainability report every year and it is available for everyone online.

Our sustainability report is prepared in accordance with the GRI Standards and Appendix 27 Environmental, Social and Governance Reporting Guide published by the Hong Kong Exchanges and Clearing Limited. We also make reference to internationally recognized guidelines and frameworks such as the United Nations Sustainable Development Goals (SDGs), the International Association of Public Transport (UITP) Sustainability Charter Reporting Guide, and ISO 26000 Guidance on Social Responsibility. We will continue to disclose our carbon footprint and key sustainability initiatives and performance in our sustainability report to maintain transparency.

In 2020, the MTR developed and published a "Climate Change Strategy", which provides a detailed account of our three-pronged approach and strategy to address climate risks, which includes the operation of a low-carbon transport network, carbon reduction (as highlighted earlier), and climate adaptation and resilience.

What other strategies do you apply to reduce your footprint?

Our main footprint comes from our rail operations, and in Hong Kong, the MTR's rail network is powered by electricity. Purchased energy is therefore the largest contributor to our carbon footprint, but compared to other road transport powered by fossil fuels, the MTR's railway operations produce much less carbon emissions. In line with the HKSAR Government's Climate Action Plan 2050, the MTR is committed to serving as the backbone of Hong Kong's low-carbon public transport network, and we are working with our utility partners on synchronized improvements.

Many of our efforts are concentrated on reducing energy consumption and improving energy efficiency across our operations.

Increasing the use of renewable energy is a key component of addressing the global climate challenge. We installed solar PV systems at two office buildings and on one of our new generation Light Rail vehicles, which provides electricity for the in-compartment lighting system. Plans are also in place to install renewable energy systems in suitable stations and depots. Hin Keng Station has been selected to pilot a solar PV system. Our target is to increase our generating capacity of renewable energy to 1 million kWh by 2023.

Besides, the Corporation also works to integrate environmentally conscious features and energy efficiency measures into new railway projects and the surrounding areas during the design, planning, and construction of new lines as far as practicable. This includes using environmentally friendly materials, installing green roof systems, and adopting low-carbon designs, all the way to targeting a minimum of BEAM Plus Gold accreditation for new stations and residential property developments.

Can you please give us some specific project examples that are directly targeting carbon emission reduction?

A number of initiatives are currently in place, e.g., a large-scale chiller replacement programme to replace over 150 chillers at MTR stations and depots is underway. It will achieve 30.4 GWh savings in energy when completed in Q1 2023. A pilot trial on AI-controlled chiller operation is also on-going, which could potentially enhance the energy efficiency of the new chillers. Regenerative braking technology has always been applied to train operations, and LED lights are used in MTR premises wherever possible. We plan to replace our existing vehicle fleet with EVs progressively and procure our first double-decker e-bus for trial.

We are also experimenting with solar power integration across our portfolio. Following the successful installation of a 189-panel solar PV system at one of our buildings in Hong Kong, another PV system comprising 296 solar panels with a capacity of 93.24 kW was installed at the MTR Headquarters in 2020. Solar panels have also been installed in one of our new generation Light Rail Vehicles to provide electricity for the in-compartment lighting system. The Corporation will review the effectiveness of this trial scheme and consider possible expansions of this initiative in the Light Rail network.

During the design, planning, and construction of new lines and buildings, we will integrate green and low-carbon features into our projects as far as practicable. For instance, our Hin Keng Station has adopted a semi-enclosed station design with generous openings on the roof, enabling optimal natural lighting and cross-ventilation. Shading made from environmentally friendly materials has been installed on the station's external wall to shield passengers from sunlight and reduce the indoor temperature. We have also installed a green roof of about 5,000 m^2 in the station and its associated structures.

How do you see the role of technology to achieve your company's decarbonization goals?

In the long run, we will continue to look for novel and scalable data-driven decarbonization solutions that enable our mobility business to become even more sustainable. For that reason, we sponsored and partnered with Carbonless Asia in launching the Carbonless Asia Challenge in 2021 to invite startups to propose technical solutions for decarbonizing urban rail transport at scale. It was a very good start, and we have got several innovative solutions from various startups, including smart management systems, which make use of big data and artificial intelligence to help reduce energy consumption and the carbon emissions associated with buildings.

The Corporation is also exploring collaboration opportunities with local tertiary institutions in Hong Kong and other cities we operate in to support smart and decarbonization initiatives and R&D projects to help the MTR meet its long-term carbon reduction goals. We will set up our first joint research laboratory with the Hong Kong University of Science and Technology on low-carbon smart city development. We will continue to work with different partners and startups to explore and identify suitable carbon-related technologies to reduce the MTR's carbon footprint.

What other factors do you think are crucial for the MTR to achieve its Net Zero targets?

Finance is definitely one, and the MTR is one of the pioneers in green finance in Hong Kong. We set up a Green Bond Framework in 2016 and further established a Green Finance Framework in 2018. In 2020, we put in place a Sustainable Finance Framework to cover a wider range of financing transactions, the proceeds of which are used for furthering the development of sustainable urban infrastructure in support of the UN SDGs.

Since 2016, the Corporation has issued green bonds to fund new/extension railway lines, railway asset replacements, energy efficiency improvements, and nature conservation projects. As of 2020, the Corporation has raised more than HK 23 billion (cca. USD 3 billion) through sustainable finance arrangements.

Is there a strategy in place to engage the general public, the users of your services, and the residents of your properties to achieve your sustainability goals?

Indeed. Apart from launching various types of initiatives in the network, the Corporation also leverages its user base and expertise to encourage the public to adopt a more environmentally conscious lifestyle. To encourage the public to prioritize low carbon transport, a "Carbon Footprint Challenge" was launched through the MTR Mobile app in 2020, providing incentives and rewards to passengers for choosing to ride on the railway network. After three successful waves of the campaign from August 2020 to September 2021, there were more than 145,000 participating registered users, saving over 61,000 tons of carbon emissions, which is equivalent to the amount of CO2 removed by over 2.6 million trees in a year.

We have also developed a brand-new mobile app called "Carbon Wallet" to promote carbon reduction actions across four lifestyle categories: recycling, dining, shopping, and mobility. Through practicing carbon-conscious behavior, users earn points based on the potential carbon emissions saved, which may in turn be redeemed for products and/or services from our participating partners. The app also provides an interactive map that allows users to find the nearest recycling points, drinking water stations, and vegetarian restaurants.

In our managed properties, we proactively seek opportunities to enhance our tenant's awareness of sustainable waste management. We work closely with the Owners' Committees and Incorporated Owners to promote waste separation programs to residents. Various types of recyclables such as used paper, used clothes, and glass bottles are collected through recycling facilities in our managed properties. Additionally, we encourage residents to join our "Central Food Waste Recycling for Improving Estate Environment" initiative to reduce their food waste through different programs, such as festive food donations and seminars. For food and beverage tenants in our shopping malls, we continue to engage them through our "MTR Malls Food Waste Reduction Pledge", with an aim to minimize waste at the source. We have also installed reverse vending machines for plastic beverage containers in some of our shopping malls to further promote waste recycling.

COP26 has just ended recently. Is there anything that the climate conference changed in your plans? Is there anything specific that you expect from the next COP?

One of the key objectives of COP26 is "Work together to deliver". We can only rise to the challenges of climate change by working together. We have to turn our ambitions into action by accelerating collaboration between governments, businesses, and civil society to deliver on our climate goals faster. We hope that the next COP can mobilize more countries, cities, and companies to accelerate their carbon reduction plans. Through collaborative efforts, we can hopefully reach net-zero carbon emissions globally by the middle of the century.

Would you have any closing remarks to the readers of this book?

Understanding that climate change poses a significant global challenge to all of us, the MTR is devoted to extending its decarbonization efforts. We are committed to sustainably managing our environmental and carbon footprint. We have concentrated our efforts and resources to provide a low-carbon transport network, improve energy efficiency, and strengthen climate adaptation and resilience measures in our operations.

To further reduce greenhouse gas emissions, we will continue to make use of new technologies and explore new solutions in collaboration with other experts or startups. We hope to achieve more — not only by adopting low carbon solutions more widely across our businesses but also by leveraging on our network to raise awareness of and promote a green lifestyle to citizens. Working together with authorities and communities in all of the cities we serve, we will strive to achieve carbon-neutral operations and make utmost efforts to avoid and limit the effects of climate change.

Endnotes

1. World Health Organization, "Air pollution", World Health Organization, https://www.who.int/health-topics/air-pollution#tab=tab_1.
2. World Health Organization, "Road traffic injuries", World Health Organization, https://www.who.int/news-room/fact-sheets/detail/road-traffic-injuries.
3. A quote attributed to Henry Ford.
4. Environmental Protection Agency, "History of reducing air pollution from transportation in the United States", Environmental Protection Agency, United States, https://www.epa.gov/transportation-air-pollution-and-climate-change/accomplishments-and-success-air-pollution-transportation.
5. According to the International Energy Agency, the growth of the world's SUV fleet caused an uptick of 0.55 gigatons of CO_2 over one decade, to 544 million tons of CO_2. See International Energy Agency, "Global SUV sales set another record in 2021, setting back efforts to reduce emissions", International Energy Agency, https://www.iea.org/commentaries/global-suv-sales-set-another-record-in-2021-setting-back-efforts-to-reduce-emissions.
6. From 2009 until 2015, Volkswagen deactivated diesel engines' emissions controls outside of laboratory reporting. As a result, global regulators were unaware that VW diesel cars emitted up to 40 times more polluting nitrogen oxides in actual driving versus what was reported. See https://en.wikipedia.org/wiki/Volkswagen_emissions_scandal.
7. The fourth and third largest chaebols, South Korean conglomerates, respectively, with activities spanning telecommunications, chemicals, electronics, and the energy sector.
8. The lingering effects are still felt today — disrupted supply chains and Chinese dominance of critical minerals processing. Building reliable supply chains and localized EV charging infrastructure across the globe will require concerted effort for many years to come.
9. Wired, "Let's count the ways e-scooters could save the city", *Wired*, https://www.wired.com/story/e-scooter-micromobility-infographics-cost-emissions/.
10. An electric moped is a lot lighter than an EV.
11. Bernard Marr, "Demand for these autonomous delivery robots is skyrocketing during this pandemic", *Forbes*, 29 May 2020, https://www.forbes.com/sites/bernardmarr/2020/05/29/demand-for-these-autonomous-delivery-robots-is-skyrocketing-during-this-pandemic/?sh=11ef8a7f7f3c.
12. Electric-style helicopters with vertical take-off and landing.

13. McKinsey & Company, "The future of mobility is at our doorstep", McKinsey Center for Future Mobility, McKinsey & Company, December 2019, https://www.mckinsey.com/~/media/McKinsey/Industries/Automotive%20and%20Assembly/Our%20Insights/The%20future%20of%20mobility%20is%20at%20our%20doorstep/The-future-of-mobility-is-at-our-doorstep.ashx.

14. COVID-19 will most likely dent international air travel and unnecessary commuting for many years to come, and the decarbonization of the energy supply will significantly reduce international shipping (some 40% of current shipping is trade in oil and gas and thermal coal).

15. Casey Crownhart, "This is what's keeping electric planes from taking off", *MIT Technology Review*, 17 August 2022, https://www.technologyreview.com/2022/08/17/1058013/electric-planes-taking-off-challenges/.

16. Approximately just under 50% of flights are under 500 miles, so this sector is meaningful.

17. Airbus and Boeing seem to disagree on the near-term commercial viability of hydrogen aircraft.

18. Worldwide biofuel production is currently in excess of 160 billion liters per year.

19. We called them "electrofuels" earlier, those made by green hydrogen and CO2 captured from the air.

20. Evelyn Cheng, "Baidu kicks off its robotaxi business, after getting the OK to charge fees in Beijing", CNBC, 25 November 2021, https://www.cnbc.com/2021/11/25/baidu-kicks-off-robotaxi-business-after-beijing-citys-fare-approval.html.

21. Logan Goldie-Scot, "A behind the scenes take on lithium-ion battery prices", BloombergNEF, 5 March 2019, https://about.bnef.com/blog/behind-scenes-take-lithium-ion-battery-prices/.

22. Ibid.

23. Battery News, "Why CATL, a China-based company, dominates electric car batteries — NYT", *Battery News*, 27 December 2021, https://batteriesnews.com/catl-china-company-electric-car-batteries/.

24. Shaandiin Cedar, "5 electric bus makers shifting into next gear", *GreenBiz*, 1 June 2021, https://www.greenbiz.com/article/5-electric-bus-makers-shifting-next-gear.

25. Brian Eckhouse and Jennifer A. Dlouhy, "Electric buses are poised to get a U.S. infrastructure boost", *Bloomberg*, 13 August 2021, https://www.

bloomberg.com/news/newsletters/2021-08-13/electric-buses-are-poised-to-get-a-u-s-infrastructure-boost.

26. Business Wire, "BYD wins the largest pure-electric bus order outside of China", *Business Wire*, 11 January 2021, https://www.businesswire.com/news/home/20210111006106/en/%C2%A0BYD-Wins-the-Largest-Pure-Electric-Bus-Order-outside-of-China.

27. The longest range Tesla ever built is equipped with a 100 KWh lithium-ion battery of ~500 km range that weighs 625 kg and has 0.40 m³ in total volume. This delivers an energy density of 160 Wh/kg against that of hydrogen fuel cells stored in high pressure tanks of 1,600 Wh/kg. This translates to 10 times more energy per kg and in a smaller volume.

28. A measure of the diversity of industry and the spread of income receipts and dependence of inputs from other countries. See The Atlas of Economic Complexity, "Country & product complexity ranking", https://atlas.cid.harvard.edu/rankings.

29. Peter Long, "Hydrogen technology cluster will help turn Karratha into industry centre of excellence", *Pilbara News*, 11 August 2021, https://www.pilbaranews.com.au/news/pilbara-news/hydrogen-technology-cluster-will-help-turn-karratha-into-industry-centre-of-excellence-ng-b881955099z.

30. Transport and Environment, "How much CO2 can electric cars really save compared to diesel and petrol cars?" Transport and Environment, https://www.transportenvironment.org/challenges/cars/lifecycle-emissions/how-clean-are-electric-cars/.

31. This depends on what fuel one is replacing.

32. Breakthrough Energy Catalysts are providing billion-dollar-scale, no-strings-attached financing to advance electrofuels for aviation use, together with green hydrogen, direct air capture, and long-duration energy storage.

33. "Drop-in fuels" are a synthetic and fully interchangeable substitute for conventional petroleum-derived hydrocarbons (gasoline, jet fuel, and diesel), meaning it does not require the adaptation of the engine, fuel system, or the fuel distribution network.

34. For example, using electricity to do the heavy lifting and extracting carbon.

35. Wikipedia, "List of countries by economic complexity", *Wikipedia*, https://en.wikipedia.org/wiki/List_of_countries_by_economic_complexity.

Section 3
NATURE

INTRODUCTION

by Roman Y. Shemakov

This section covers our lands, oceans, and carbon. From the emergence of the earliest homosapians and the infrastructures of Gubeklitepes 13,000 years ago, we have been terraforming our environment: caricaturing natural material into tools to domesticate animals, tinker with plant biology, and accumulate metabolic security. The combination of the three has left an unmistakable human mark on the surface of the planet. Dams brought life to deserts, cities razed forests, and locomotion has moved million of years of carbon within the Earth's surface back into the air. Much of the most transformative environmental change has occurred within the last 200 years. A slew of earth scientists and geologists have noted that the impact of humanity on the planet has been significant enough that it constitutes its own geological age: the Anthropocene.

The solution to this acceleration must be as artificial as its cause. We cannot escape to the land, maintain the status quo, and abdicate our influence on the natural world. It is vital to acknowledge that while we have had a profound impact on our environment, unilateral national and company decisions have inadvertently disturbed the million-year-old geophysical rhythms of the planet. Our knowledge about these rhythms remains limited. Nonetheless, humanity must embrace its capacity to transform the environment if we are to recreate a viable planetary equilibrium.

This section outlines the problems present in our relationship to the planetary metabolic cycle and the species-scale solutions necessary to return the planet to a predictable geochemistry. We must bury the carbon in the sky, return minerals to the soil, and remove trash from the oceans. The same multinational efforts that brought humanity to outer space and interconnected the world via railroads are now needed to reverse the environmental course. Such global problems require global solutions that should be based on insurance, distribution of risk, and equitable sharing of benefits.

In "Food & Agriculture", Alexandra Tracy investigates the steps necessary to decarbonize Asia's food chain. The sector presents a variety of challenges. Considering that agriculture and deforestation account for 20% of the greenhouse gas emissions from Asia, a continent where

500 million people are already malnourished, uncertain climate variations are threatening the food supply of the most populous continent on the planet. With accelerating urbanization and shrinking agricultural capacity, reducing vulnerabilities and increasing resilience is the most pressing concern right now. Fortunately, a myriad of technologies, i.e., drones, drip irrigation, cold storage, transportation, and blockchain, is offering the resources to strengthen the precarious food system. Carbon credits, insurance, and shared service platforms have the potential to support 100 million smallholders in producing 80% of all the food consumed on the continent.

In "Forestry & Oceans", Bill Kentrup looks at the tools we must employ to mend the natural environments. Our oceans trap nearly 40,000 gigatons of CO_2, and another 4,000 gigatons are held by forests, vegetation, and permafrost. To put that in context, the burning of fossil fuels contributed 34.8 gigatons of CO_2 last year. The chapter outlines the changes in land use, carbon finance, cap and trade, as well as monitoring and the accounting necessary to secure the planet's most important natural resources.

In the book's final chapter ("Sequestration"), Moon K. Kim analyzes the most important factor necessary to maintain a livable ecology: removing excess carbon from the planet's atmosphere en masse. Hypothetically, if the planet is a bathtub and CO_2 is water, one must not only turn off the tap but also clear the drain. On a planetary scale, scientists estimate that more than 10 gigatons of CO_2 must be removed annually from the atmosphere by 2050. Large-scale sequestration projects must grapple with two central challenges: capturing and storing CO_2. Beyond natural methods like restoring forests and wetlands, the global effort must become aggressively artificial. The chapter focuses on the methods needed to make direct air capture, storage, and carbon transformation a ubiquitous network of global infrastructure in the 21st century.

Chapter 7
FOOD & AGRICULTURE

by Alexandra Tracy
and Archawat Chareonsilp

ANALYSIS

by Alexandra Tracy

Decarbonizing the food chain in Asia presents a range of particularly complex challenges. Agriculture is essential to livelihoods in the region's emerging markets, where more than 100 million smallholders produce over 80% of the food consumed.[1] In an area where nearly 500 million people are still malnourished,[2] food security remains the priority for many policymakers.

But farmers in the region are already being hit by the impacts of climate change, and the sector is a major contributor to greenhouse gas (GHG) emissions — far more than in most Western economies, as can be seen in Figure 7.1. Agriculture and deforestation, through burning and clearing, account for over 20% of emissions in Asia,[3] including 40% of methane emissions (largely from livestock farming). In addition, off-farm activities contribute to a growing share of food system emissions across the entire production, transportation, and consumption chain.

Climate change is also putting increasing pressure on the availability of water supply in many countries in Asia, where the agriculture sector

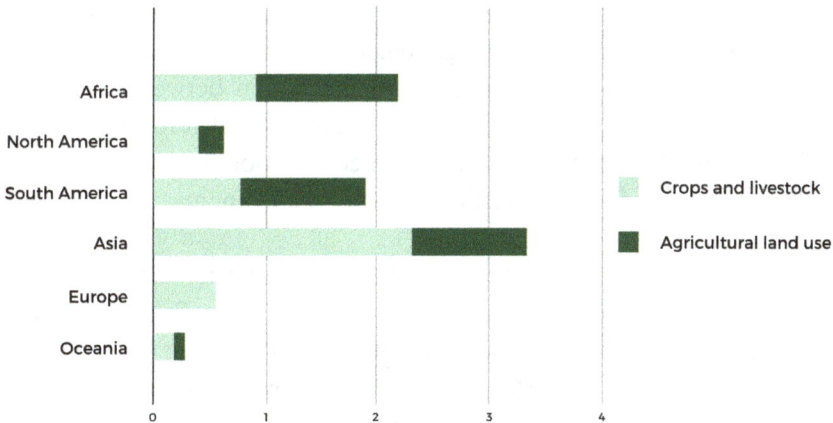

Figure 7.1. Emissions from crops, livestock, and agricultural land use.
Source: FAO.

235

currently consumes more than 80% of all water resources.[4] It is likely that usage in the sector will increase in emerging markets over the short to medium term, complicated by strong competition from households and industry for scarce water resources. Future scenarios suggest that many of the region's transboundary river basins will be stressed or highly stressed.[5]

While these environmental challenges remain significant, the agriculture sector in Asia is in the early stages of a sweeping transformation, which will combine mechanization and automation with better inputs and greater availability of information about production techniques and market trends. With support from government, academia, and international bodies, access to technology is gradually being extended to smallholders and poor rural communities, which will both increase their productivity and have a material impact on emissions.

PAST

As Asian economies emerged from the post-war period, poverty alleviation was the immediate priority, with agriculture receiving the bulk of supportive policy initiatives, and rapid demographic change leading to food security concerns. However, new practices, such as high-yielding seed varieties and access to irrigation, fertilizers, and pesticides, dramatically expanded crop production in the region between 1970 and 1990. With the introduction of extensive agricultural reforms in 1978, China began rebuilding from its farm collectivization program and introducing widespread mechanization, which greatly increased wheat and rice volumes.[6]

However, productivity improvements have plateaued, and after decades of steady decline, the number of people who suffer from undernourishment began to slowly increase again in 2015.[7] Reversing this trend will be essential to reducing rural poverty and feeding an Asian population, which could rise to 4.9 billion by 2030.[8]

Growing Incomes and Urbanization Driving Demand for Food

Asia's middle class is growing at an unprecedented rate and may reach over 2.2 billion people by 2030.[9] By the same date, as we explain in Chapter 5, 555 million people are likely to have moved into cities, where they will create more than 85% of the gross domestic product.[10] This, in turn, will expand consumer spending on food, which may double to more than USD 8 trillion by 2030. India and Southeast Asia are expected to account for the greatest increases, growing at a compound annual rate of 5.3% and 4.7%, respectively. China will remain the largest market overall.[11]

There will also be considerable demand for new types of food, especially meat and dairy products. For example, per capita beef consumption alone in China could increase by over 40% between 2010 and 2030,[12] while the market for meat products in Southeast Asia is estimated to grow at over 5% a year to nearly USD 120 billion.[13]

Shrinking Agricultural Capacity

With 20 megacities expected by 2025,[14] urban areas in Asia are rapidly encroaching on agricultural land. China alone may account for a quarter of total global cropland loss by 2030; Pakistan, Vietnam, and Indonesia are also major potential losers.[15]

As people migrate into cities, workers are lost to the agricultural sector. In Indonesia, for example, urbanization could result in about 8 million fewer farmers by 2030.[16] As fertile land is lost and the numbers of productive workers shrink, previously untouched areas have to be cleared and brought into cultivation.

Government Interventions Create Problems

Governments have responded to the challenge of food security with a number of mechanisms to provide support to farmers. In addition to tariffs and quotas to protect domestic markets, about USD 540 billion is paid in agricultural subsidies worldwide each year.[17] Incentives to increase production create enormous pressure on natural resources, especially water.

More than 450 million people in Asia already experience very high water stress, particularly in India, China, Indonesia, and Pakistan. Climate change is expected to exacerbate shortages by changing rainfall patterns. At the same time, farmers face increasing competition for water from industrial and residential users.[18]

Government interventions to increase food yields are also pushing up emissions. For example, subsidies for chemical fertilizers have led to considerable overuse in many parts of Asia, generating substantial emissions of nitrous oxide, a GHG that is >300 times more potent than carbon dioxide.[19] Synthetic fertilizers alone account for 13% of agricultural emissions.[20] In China, traditionally the largest user in the region, the government has been scaling back subsidies, but India recently announced an extra USD 8.71 billion for the current fiscal year.[21]

Big Food Supply Chain Dominance

Over the past three decades, the largest agribusiness and food companies have become more global and created complex production, manufacturing, and supply chain networks, giving them enormous influence over industry practices and inputs. Environmental concerns are becoming more important to their operations, but a recent study found that over 70% of the world's 60 largest publicly listed meat, dairy, and aquaculture producers are failing to manage climate risk.[22] According to the Farm Animal Investment Risk and Return (FAIRR): "If global animal agriculture was a country, it would be the second-highest emitter of GHGs", as food companies such as Nestle and Macdonalds outsource huge parts of production overseas to suppliers with very limited emissions reduction strategies.[23]

PRESENT

Recent extreme weather events around the region have highlighted the vulnerability of the farming sector to potential climate change. Flooding in China's Henan province in 2021 impacted around 2.4 million acres of cropland, as well as food processing facilities.[24] India[25] and the Philippines[26] also experienced torrential rainfall, causing mass evacuations and crop damage. At the same time, a severe drought in Central Asia caused shortages of water for irrigation and huge livestock losses.[27]

Against this backdrop, agribusiness and food companies are responding to increasing pressure from governments as well as investors and consumers to improve environmental performance, reduce inefficient production, and tackle emissions. Technology is beginning to be more widely utilized both at the farm level and across the supply chain.

Technology At the Farm Level

Greater use of agricultural machinery reduces the need for human labor, while the increasing adoption of computer systems, electronics, and data management on Asian farms will create enormous cost savings and efficiencies over the medium term.

Precision Agriculture

Precision agriculture, using high technology sensors and analytical tools, improves crop yields and assists farming management decisions. It uses data to monitor crop status and soil condition and manage processes, such as irrigation, planting, and use of fertilizers and pesticides, more efficiently. In much of Asia, the technologies are unaffordable for most farmers but are becoming more widespread in the developed markets.

In South Korea, for example, the Rural Development Administration is promoting "digital agriculture" for the farming industry and encouraging efforts to implement high-efficiency precision farming, as well as providing support for young farmers to utilize these technologies.[28] Similarly, in Japan, the government is pushing for agricultural reforms involving big data, the internet of things, and artificial intelligence (AI). The Ministry of Agriculture, Forestry and Fisheries published a roadmap for business expansion into smart farming technologies and services to be achieved by 2025 and has established over 120 demonstration projects.[29]

Drone Technology

Drones play a key role in precision agriculture, both for gathering data and replacing manual labor, an increasingly important consideration for a region with a shrinking rural population. Agriculture drones can spray 40–60% faster than manual spraying, creating savings in the use of chemicals of 30–50% and up to 90% less water use.[30] In China, annual sales of agricultural drones have grown from fewer than 1,000 in 2017 to 15,300 in 2020,[31] reducing the need for tractors and mechanical sprayers. In the Heilongjiang province, for example, more than 80% of rice was already being sprayed by drones in 2019.[32]

Drip Irrigation

As urban diets include more meat, a great deal more water will be required for its production. A large proportion of irrigated farmlands in the region is served by pumping groundwater, which is extremely energy-intensive. In India, lifting water for irrigation alone is estimated to contribute as much as 11% of the country's emissions.[33]

Drip irrigation feeds the plant instead of the soil, delivering water and liquid fertilizers straight to the roots, which greatly lowers water consumption. When combined with solar-powered pumping equipment, it also reduces emissions in rural areas. With financial support from the Ministry of New and Renewable Energy, the state of Rajasthan in India has become the country's solar water pump leader, allowing farmers to replace diesel and giving them higher yields and multiple harvests in a year.[34]

Improved Cultivation Techniques

Rice cultivation generates around 50% of all crop-related emissions[35] due to the age-old practice of flooding paddy fields to prevent weeds from growing, which also produces methane. Addressing this problem needs to be a priority, as rice (unlike meat) is the staple food of hundreds of millions of people in Asia. The Sustainable Rice Platform, based in Bangkok, is working with farmers in Thailand, Vietnam, and elsewhere to enable them to grow the crop more sustainably.[36] Rice that complies with its standards is eligible for certification, which is supported by global buyers such as Olam International in Singapore.[37]

Regenerative Agriculture

Altering farming and grazing practices can help to address climate change by rebuilding soil organic matter, improving watersheds, and restoring degraded soil biodiversity, which can reduce emissions and create new carbon sinks. The regenerative agriculture movement originated in the United States, where some food retailers have been very supportive. Walmart, for example, has committed to protect and restore at least 50 million acres of land by 2030.[38]

In emerging Asia, the prevalence of smallholders on very limited plots with uncertain land tenure makes regenerative agriculture — which typically requires scale — more challenging. In many cases, capacity constraints in local government agencies and disconnects between development, land use, and environment departments mean that the potential for public support for farmers is limited. International agencies such as the UN Habitat are making efforts to work with policymakers to address these issues, while a number of local pilot projects have been established, such as a partnership between agrichemical company Syngenta and The Nature Conservancy in China[39] and the Harmless Harvest's Regenerative Coconuts Agriculture Project in Thailand.[40]

Urgent Need to Reduce Food Loss and Wastage

Every year, approximately 690 million tonnes of food is lost and wasted in Asia,[41] meaning more production is required to compensate for this. The majority of losses occur in emerging markets, where poor storage facilities and inadequate transport infrastructure mean that a large amount of food is wasted after harvest. In the richer, more urban areas, growing volumes of food waste at the retail and consumer levels have to be disposed of by incineration or in landfill, both creating emissions.

Cold Storage and Transportation

In India, only about 10% of perishable food is kept in cold storage systems, leading to a loss of up to 30% of fruits and vegetables.[42] In response, the government launched the India Cooling Action Plan to provide better access to sustainable cooling technology.[43] In neighboring Bangladesh, Solar E Technology provides affordable, solar-based micro-cold storage that can replace traditional ammonia-based stores.[44]

Automated freezer storage, using robots to stack and retrieve food products efficiently, reduces food wasted through human error or delays. For example, SK Cold Chain in Malaysia offers fully automated cold storage facilities and warehouse management systems.[45]

Urban Retail

Food packaging in a typical Asian supermarket, where the onus is on hygiene and appearance, contributes materially to the overall emissions in the supply chain. There has been recent pressure to move away from single-use plastics, and some countries, including China, are considering allowing recycled plastic materials.[46] Innovative packaging can also increase the shelf life of fresh food and reduce wastage. In Singapore, for example, the government's Agency for Science, Technology and Research (A*STAR), together with a local company, Dou Yee Enterprises, have developed a new material for food packaging that can extend the shelf life of food by at least 50%.[47]

A wide range of alternative products is also emerging in Asian markets, in particular, new protein sources that can replace meat while meeting specific regional tastes. For example, Hong Kong-based Right Treat has produced a pork substitute, and Zhenmeat in Beijing offers a range of alternatives catered to suit traditional Chinese cuisine.[48] (See "Lifestyle & Consumption".)

End User Demand Management

Some governments in Asia are using regulation and taxation to reduce food waste. For example, in several cities in South Korea, authorities have introduced a policy requiring households to pay for recycling services according to the quantity of food waste disposed of, which has reduced total volume by over 10%.[49] Parts of China have similar schemes for both households and restaurants.

Restaurants in the region are beginning to use inventory management systems to track and reduce food waste. A British startup, Winnow Solutions, with offices in Singapore and Shanghai, uses smart meters to track everything thrown away in a restaurant kitchen and allows staff to alter menus and portion sizes accordingly. Users have been able to achieve a reduction in food waste of up to 50%.[50]

FUTURE

There is an existing base of agricultural technologies that will be scaled up in the region over the next decade, with innovation in local markets creating disruption and opportunities for startups. Mobilizing more investment into the sector will be essential for maintaining technology advancement, and incremental change will be driven by technology and financing.

Urban Agriculture

As Asian populations increasingly move into cities, urban and peri-urban areas will have to play a much larger role in food security, increasing local supply and lowering costs and emissions through reduced transportation and storage. Ranging from relatively low-tech community gardens to highly sophisticated plant factories, the opportunity for urban agriculture in the region is estimated to be around USD 20 billion by 2030.[51]

Vertical Farming

Indoor vertical farming, incorporating hydroponic and aeroponic technologies and controlled lighting and integrated sectors, overcomes weather constraints and allows crops to be grown more quickly on a much smaller land area with significantly greater efficiency. For example, in Japan, indoor farms have reduced water usage by 99% compared with land farms.[52]

In Singapore, startup Sustenir Agriculture retrofits existing buildings in the city to create a controlled environment for growing and is able to sell vegetables at 30% of the cost of imported alternatives.[53] Another local company, Apollo Aquaculture Group, has created a vertical fish farming facility that produces six times more than a traditional aquaculture project.[54]

As these techniques mature, vertical farming in the region is likely to attract larger-scale investment, with companies such as SPREAD[55] and Mirai[56] in Japan, and Noonty Greenhouse Co Ltd in China, already seeing huge growth in demand.[57] The Asia-Pacific market, as a whole, is expected to expand by 24% a year to 2026.[58]

Data to Enable Better Forecasting and Traceability

The use of data gathered from remote sensors and drones is becoming essential to precision agriculture and indoor farming practices. In addition, some companies in the region are exploring the use of AI and blockchain technology for demand management and to improve the traceability of their products.

Artificial Intelligence for Improved Forecasting

The Japan Weather Association collaborated with food producers such as Sagamiya Foods, which makes tofu, and sauce company, Mizkan Holdings, to develop an AI system to forecast food demand based on weather information and sales data. The system aims to help companies scale back redundant production and cut food inventory losses.[59] At the consumer level, a startup in Singapore, Lumitics, uses AI to help hotels and airlines to understand how much food waste they are generating and adjust their planning decisions about what produce to buy and serve.[60]

Blockchain Platforms Offer Traceability

Blockchain could radically transform agricultural supply chains in Asia by providing visibility on quality and provenance, creating consumer trust, and making it easier for producers to access new markets. While usage is not widespread among small farmers due to cost, there is huge potential for improving transparency and creating direct payment options in emerging and more remote agricultural markets.

For example, WOWTRACE, a startup in Vietnam, provides data on the supply chain from cocoa beans to a final product in the premium chocolate market (see Figure 7.2).[61] In the Pacific islands, a collaboration with the UN in Jiwaka in Papua New Guinea is enabling pork farmers to prove that their livestock meets international standards,[62] while the Blockchain Supply Chain Traceability Project, coordinated by the World Wildlife Fund in New Zealand, tracks tuna in the Pacific from vessel to supermarket.[63]

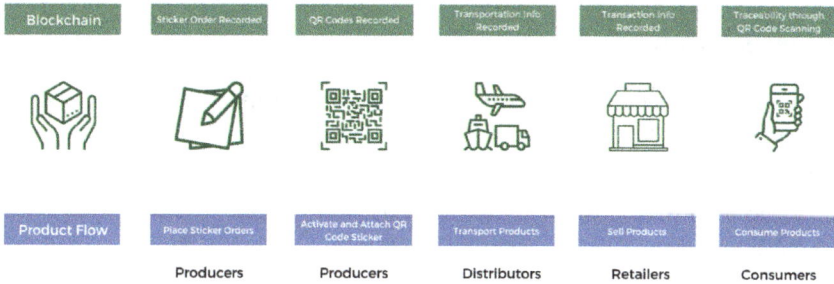

| Blockchain | Sticker Order Recorded | QR Codes Recorded | Transportation Info Recorded | Transaction Info Recorded | Traceability through QR Code Scanning |

| Product Flow | Place Sticker Orders | Activate and Attach QR Code Sticker | Transport Products | Sell Products | Consume Products |
| | Producers | Producers | Distributors | Retailers | Consumers |

Figure 7.2. Traceability supply chain.

Financing

More institutional capital is being directed into climate-friendly food production in Asian markets, much of it catalyzed by multilateral and development banks or "patient" philanthropic funding. Providers of microfinance have traditionally been active in the sector, and dedicated funds have recently been launched by major institutions such as HSBC,[64] Rabobank,[65] and Lombard Odier.[66] At a local level, new digital business models are also providing improved access to financing.

Fintech and Shared Service Platforms

In the Philippines and Indonesia, digital peer-to-peer lending platforms, which allow urban professionals to make small loans to multiple farmers in rural areas, are experiencing double-digit growth.[67] Combining digital payment and shared service platforms (where farmers pay to use equipment only when they need it) gives smallholders access to essential machinery and resources. For example, Tun Yat, a startup in Burma, allows farmers to rent tractors and harvesters by the day through a mobile app.[68] In India, Ergos provides warehousing to rural farmers, enabling them to store and sell their produce in the off-season for higher prices. Farmers can use digital warehouse receipts as collateral to secure short-term funding until their crops are sold.[69]

Insurance Coverage to Support Investment

Although insurance can be essential to building climate resilience, coverage has typically been extremely poor in Asian emerging markets. Affordable parametric insurance, which pays out automatically in response to predefined incidents such as wind speed or rainfall, is now available in several parts of the region. For example, CARD Pioneer Microinsurance Inc., with support from the International Finance Corporation, developed a low-cost product to protect rice and corn farmers against losses caused by typhoons.[70] The first such scheme in the Pacific Islands was launched this year in Fiji by FijiCare and Sun Insurance, with UN backing, covering smallholders, fishermen, and market vendors.[71]

Carbon Credits Provide New Revenue Stream

More efficient farming practices, which reduce the need to develop virgin land, and climate-friendly land management practices enabling soil carbon sequestration can generate carbon credits to be sold in the voluntary markets. While agricultural credits are still a relatively untested product, especially in Asia, the potential exists to create significant new revenue streams. In a landmark deal earlier this year, for example, Australian-owned Wilmot Cattle announced the sale of AUD 500,000 worth of soil carbon credits to Microsoft.[72] Japan's Sumitomo Corporation recently signed an agreement with American company Indigo Ag, a leader in the agriculture carbon credits business, to explore opportunities for soil carbon projects in Asia.[73]

CONCLUSION

To meet the twin challenges of decarbonization and long-term climate resilience, the agriculture sector in Asia needs to embrace an extensive digital and technology transformation. In most countries in the region, advances in productivity to date have come from mechanization and better inputs. Going forward, more sophisticated digital tools, such as connected sensors, analytics, and AI, will deliver future gains, as well as improving the management of natural resources and ensuring emissions reductions.

This transformation will be enabled by connectivity. While some parts of Asia still lack the necessary infrastructure, coverage is increasing rapidly. By 2030, it is expected that advanced connectivity infrastructure of some type will cover around 80% of rural areas.[74] At the same time, as the costs of hardware and digital devices fall rapidly, even smallholders in the poorest regions will be able to access multiple data sets as well as the analytical tools required to extract real value.

Greater connectivity is also likely to change the agricultural ecosystem in Asia and create opportunities for new players to enter the sector, ranging from telecommunications and network providers supplying the infrastructure to enable digital applications on farms, to agritech companies offering technology and data solutions to farmers.

STARTUP CASE STUDY
Ricult

by Archawat Chareonsilp

Industrial agriculture was introduced in the early 19th century to improve farm productivity and feed the world's growing population. Since then, food has been mass-produced and available at an affordable price. However, industrial agriculture was achieved at the expense of the environment as forests were cleared for farmlands laden with chemicals. As a result, more and more GHG was released into the atmosphere due to industrial farming activities. In 2018, agriculture and related land-use emissions accounted for **17%** of global GHG emissions from all sectors.

Can the agriculture sector continue to feed the world's growing population without destroying the environment? YES, it can, but a systematic value chain change must take place to allow different kinds of agriculture and food production to flourish. More researches are becoming available, demonstrating that when sustainable farming practices are introduced, the same level, if not better, of farming productivity can be achieved.

Even though small and slow, such positive changes are happening across the geographies. Conservation farming and regenerative agriculture practices have been gaining momentum in the last decades. The 17% GHG emissions from the agriculture sector measured in 2018 is already lower than the figure in the 2000s, which measured at 20%.

The challenge facing the agriculture and food sector is promoting and scaling the new farming methods faster, better, and cheaper.

This section shows how Ricult, a startup, engages farmers in Thailand to start the conservation and regenerative farming journey. Also featured in this section is the global agriculture company Syngenta and how they support the regenerative movement with knowledge, innovations, and technologies. Lastly, we will look at how one of the world's biggest consumer brands, Coca-Cola, empowers consumers to take action on GHG and climate change.

Growing Concerns on the Effect of Climate Change on the Thai Agriculture Sector

According to *Climate Risk Country Profile* published by the World Bank Group and the Asian Development Bank, Thailand is among the top ten most flood-affected countries in the world. Thailand's agriculture sector could be significantly affected by floods, extreme weather events, and the rising temperature, especially in the south.

Meanwhile, Thailand has a significant footprint on the world's food and agriculture export. The Kingdom's agricultural prowess earns her the top rice, sugar, and natural rubber exportation rankings. According to the United Nations (UN), 40% of Thai labor is employed in the agriculture sector; however, the sector only contributed 9% of the country's GDP. Moreover, smallholder farmers occupy nearly 80% of the total farmland; their activities significantly impact land use and the overall level of CO_2 emission.

With an average income of below USD 200 per month, most Thai farmers still live under the poverty line. Therefore, to empower smallholder farmers toward a more sustainable agriculture practice, in general, and decarbonization, in particular, is a diligent balancing act between profit and planet.

Sowing the Seeds of Regenerative Technology

Founded in 2016 by four MIT graduates, Ricult, an AI-powered agritech startup, deploys deep tech, AI, and machine learning to help process and manage agricultural work accurately and efficiently. Ricult utilizes a wide range of technology solutions such as a micro-climate weather model, satellite imagery, and crop modeling to develop a data platform to help farmers and agribusinesses optimize their value chain potential. For example, farmers can increase farm productivity, banks can effectively offer smallholder farmers affordable loans, and insurers can underwrite the future crop loss liabilities more accurately.

Currently, over 500,000 farmers have subscribed to Ricult's platform.

The startup's early success with smallholder farmers was due primarily to providing easy access to technology while boosting farmers' profit through more accurate weather data allowing the optimum timing from seeding to harvesting. In addition, more robust seeds helped farmers use less water and fertilizer. As a result, healthier crops are better protected from pests and disease, lessening the dependency on chemicals and pesticides. While these added to a better bottom line, they indirectly helped improve soil health and the optimum land use that protected the much-needed carbon sink in the farming communities. Ricult also established a zero-waste program, transforming agricultural waste into a renewable energy source for the nearby bio-mass power plant.

The Missing Heartware

It appeared that a lot could be accomplished by introducing smart technologies. However, based on Ricult's intensive interaction with smallholder farmers, achieving the delicate balance between profit and planet is no simple task.

When making choices on the farming activities, most smallholder farmers in Thailand prioritize profit over the planet. Therefore, terms such as Climate Change and Decarbonization are almost irrelevant to farmers living below the poverty line. Empowering farmers to become active climate actors, Ricult is improving its data platform to integrate the related environmental impact data with the current smart farming data.

Unlearning and Relearning Agriculture

Ricult believes that one of the critical factors determining whether smallholder farmers will subscribe to regenerative farming practices is the ability to provide actionable data and clear incentives. While the standard protocol to measure CO_2 emissions from the farming activities is still lagging, Ricult is developing a new analytics dashboard to feature an additional set of data, initially highlighting soil health, activity-based CO_2 emission, and plant and tree CO_2 sequestering data. Such a dashboard will constantly nudge smallholder farmers to pay attention to soil quality and biodiversity while helping key stakeholders maintain visibility on CO_2 emissions. Closing the data loop, the new dashboard will also track the potential CO_2 sequestered by plants, trees, and soil. Ricult hopes that when cultivated and certified, this data will serve as an additional source of income for small-holder farmers when a matured agricultural carbon market is established, hence an incentive for the farmers to adopt regenerative farming practices.

Knowledge is another factor determining whether smallholder farmers will subscribe to regenerative farming practices. More chemicals and fertilizer do not always mean better quality and yields. There is a better way to maintain the farm. Knowledge is a critical connection that can hold together profit and the planet. Smallholder farmers' lives are rooted to the land in more than one way — it is a loving home, a root of their culture, a business, and a long-term asset for generations to come. It is also true that their view of their land is often through the lens of profit. Ricult is striving to introduce the lens of the planet and connect the two dots. Fertile land and smart farming deliver better quality and yields while optimizing land use, preventing forest encroachment, and sequestering more CO_2 that help reduce the global temperature in general and provide a better livelihood for their families in particular.

Better quality and yield, on the other hand, do not always imply profit. By shifting to regenerative agriculture, Ricult wants to ensure that the markets can offer better pricing for these products. By collaborating with various food processors and government units to promote more sustainable farming practices, Ricult provides greater opportunity for smallholder farmers who are committed to both profit and the planet and recognize their role and contribution toward climate action in general and CO_2 emission reduction in particular.

More About Ricult

In 2021, Ricult successfully closed the Pre-A fundraising round with a new cash injection of USD 3.5 million from Sojitz Corporation, one of the leading conglomerates from Japan, and elea Foundation, an impact investment firm from Switzerland. This is in addition to the earlier investment from Bualuang Ventures Company Limited, a subsidiary of Bangkok Bank Public Company Limited, and Krungsri Finnovate Company Limited, a subsidiary of Bank of Ayudhya Public Company Limited, bringing the total amount raised to close to USD 6 million. Ricult will raise another funding round in 2022 (Series A).

CORPORATE CASE STUDY
Syngenta

by Archawat Chareonsilp

Agriculture and Climate Change

According to the Food and Agriculture Organization (FAO), the total CO2 emission from the related agriculture and land use was measured at 9.3 billion tons in 2018, representing 17% of the total global emissions, down from 24% in the 2000s. However, it is worth noting that this slight decrease in 2018 resulted from a reduction in CO2 emission from deforestation and a relatively greater emission release rate from other industries. On the other hand, the emission from farm activities continued to increase.

Climate Change is Everyone's Business

The food and agriculture sector remains one of the primary sources of CO2 emission. Achieving CO2 reduction across the industry will significantly contribute to keeping the global temperature increase below 2°C. Global agriculture companies must play an active role and contribute to this goal. Responding to the question during the World AgriTech Innovation Summit in 2021, Erik Fyrwald, CEO of Syngenta, shared his vision and commitment to the cause.

"While global populations continue to increase, food production systems need to produce more, more efficiently, in ways that protect the environment. This is both imperative and — using sustainable agricultural management practices — possible. With the right knowledge, practices, and technologies, farmers can play an important role in tackling climate change, helping not only to mitigate carbon emissions from agriculture but also sequestering more carbon into the soil. By supporting this, our food and agriculture systems could remove more greenhouse gases from the atmosphere than they place into it. This requires partnerships between governments, the private sector, and civil society. It also requires setting ambitious goals that take an outcomes-focused approach and that create the space for companies to innovate to help achieve those outcomes."

The Syngenta Group is a leading global agricultural science and technology provider, particularly seeds and crop protection products, with its headquarters in Basel, Switzerland.

Commitment to Growing Well and Good

Given Syngenta's broad portfolio of products, many of which are pesticides, the company, from time to time, is being criticized for causing harm to human health and the environment. For example, when misused, pesticides could poison farmers and damage the much-needed soil biodiversity, hindering the soil's ability to absorb CO_2 and causing it to be released back into the atmosphere. Today, Syngenta is working with farmers worldwide to address this paradox as pesticides continue to play a vital role in protecting crops and ensuring a high level of yields necessary to feed our growing population.

Launched in 2013 and updated in 2020, Syngenta's Good Growth Plan is among the company's core strategies that ensure the long-term viability of the business and the planet's ecosystem. According to Syngenta's 2019 Good Growth Plan Progress Report, the company is making progress toward its six commitments:

Make crops more efficient: Targeted to increase the average productivity of the major crops by 20% without using more land and water. As a result, by 2019, Syngenta managed to achieve an 18.8% increase in crop productivity.

Rescue more farmland: Syngenta is committed to rescuing 10 million hectares of degraded farmland. By 2019, programs focusing on improving soil health were deployed, and 14.1 million hectares of degraded farmland across the globe were enhanced, including those in China and the Philippines.

Help biodiversity flourish: Syngenta is committed to enhancing biodiversity on 5 million hectares of farmland. By 2019, 8.2 million hectares of farmland benefited from the biodiversity project.

Empower smallholders: Syngenta is committed to reaching 20 million farmers and enabling them to increase productivity by 50%. By 2019, Syngenta reached 20.5 million farmers and helped increase farmers' productivity by 28.5%

Help people stay safe: Syngenta is committed to training 20 million farmworkers concerning safety practices, especially in developing countries. By 2019, 42.4 million farmworkers were trained.

Look after every worker: Syngenta is committed to a fair labor practice throughout its supply chain.

From Commitment to Innovations and Actions in Asia

Data, technologies, and innovations are the bedrock that supports the commitments above. Accordingly, Syngenta is investing USD 2 billion for innovation to tackle Climate Change. From this investment, Syngenta expects to deliver two sustainable agriculture tech breakthroughs each year starting in 2019.

In February 2021, in partnership with the Nanjing provincial government, Syngenta launched the National Center for Agri-Technology in China. The Center is the Syngenta Group's most advanced global innovation research and development (R&D) center and agro-tech science hub. Syngenta invested USD 230 million to build the state-of-the-art crop protection innovation program designed to support Chinese farmers to handle extreme weather events better as well as improve food sustainability and security situations in China and Asia.

Furthermore, conservation agriculture practices were introduced to help improve soil health, reduce GHG emissions, and increase carbon sequestration in soil. For example, Syngenta launched a program to enhance soil health for 2 million hectares of degraded farmland in China. The conservation practices deployed in China, among other things, were the conservation tillage and cover cropping. These practices ensured as little as possible soil disturbance that could damage soil fertility as well as biodiversity. In addition, farmers in China are encouraged to incorporate straw and crop residues into the soil to improve organic matter content.

These methods help reduce GHG emissions and promote carbon sequestration, turning the land into a net carbon sink.

To further help biodiversity flourish and improve carbon footprints, marginal field margins management was introduced. Farmers are encouraged to utilize their land's margin or less productive areas to enhance local biodiversity. These marginal spaces do not see much of the farming activities, hence is an ideal space for wildlife to thrive, absorb surrounding air pollution, and sequester.

By applying advanced genetics, pest-resistance traits, and cutting-edge crop protection solutions in the reference farms, Syngenta sees good progress in crop productivity. Based on data collected over time, GHG emission efficiency in the reference farms increased by 36.7% compared to other farms. This information is shared, used, and appreciated across Syngenrta's network.

In 2019, Syngenta partnered with The Nature Conservancy (TNC) to scale up sustainable agricultural practices by leveraging Syngenta's research and development capabilities and TNC's scientific and conservation expertise. The partnership enabled Syngenta to expand its global efforts on sustainable agriculture, allowing the company to apply and test innovative new techniques to enhance soil health, protect natural habitats, and improve carbon sequestration in agriculture across the geographies. For example, Syngenta is working with TNC to improve soil health and productivity in a potato-growing region in China. Testing and research are being conducted to understand the consequence of continuous cropping of potatoes and to explore a science-based, sustainable crop rotation for the region.

In addition to The Nature Conservancy, Syngenta is also engaging with other partners such as The Climate Smart Agriculture Project, The Business Council For Sustainable Development, The World Economic Forum, and the Global Alliance For Climate Smart Agriculture to empower the food and agriculture systems to achieve carbon-neutrality and meet the 2°C target set by the Paris Agreement.

In 2019, Syngenta emitted 7.3 Mt of CO_2 annually, 88% of which is represented by its supply chain. As one of the world's leading agricultural science and technology companies, Syngenta's manufacturing and operations contribute to the company's overall GHG emissions. As a result, Syngenta is committed to reducing the GHG emissions intensity from operations by 50% by 2030. In pursuing this reduction commitment, Syngenta developed targets and a timeline validated by the Science Based Targets Initiative (SBTi).

Author's Note: The Future of Growth

Growing more food for the growing population is necessary to ensure economic and social security. The conservation and regenerative agriculture movement represent the growing opportunities for the new age of agriculture in Asia. The key is to build a regenerative agribusiness that is profitable for everyone — the company, the farmers, the community, and the consumers. To achieve this, Asian agribusiness companies can play a role in promoting the four elements contributing to the farming successes:

- *Knowledge about conservation and regenerative practices. Demystify the belief that they are less productive and less profitable by providing tangible and actionable data.*

- *Investment in product R&D that supports commercial scaling of the demonstration farms or reference farms. Farmers will require additional financial support to transition from the current industrial agriculture to conservation and regenerative agriculture.*

- *Create a network and market for conservation and regenerative agriculture products.*

- *Advocate for suitable government policies supporting the transition from the current industrial agriculture to conservation and regenerative agriculture.*

INTERVIEW
Dr. Casper Durandt
Sustainability Director of Asia
and the South Pacific,
The Coca-Cola Company

by Archawat Chareonsilp

The Coca-Cola Company is 135 years "young". With over 2 billion beverages served daily, pre-COVID-19, Coca-Cola is one of the most engaged brands in the world. The secret to Coca-Cola's success and longevity is its ability to stay ever-relevant, and that is not limited to only providing refreshing beverages and inspiring marketing. Being ever-relevant also means that the company has the right positions and actions on critical matters, including climate change.

The continued success and longevity of Coca-Cola in the 21st century will be defined, among other things, by how well it can transform and decarbonize to help limit the global temperature increase within 1.5°C by 2100. Furthermore, being one of the most engaged brands, it is imperative that Coca-Cola empowers its customers and consumers as it embarks on its decarbonization journey.

Kantar Public recently conducted a survey titled "Are citizens ready to change their lifestyle to address the climate crisis?" According to the survey, citizens in nine developed countries, including Singapore, are highly aware and concerned about climate change. Furthermore, their perception is that they are personally committed to preserving the environment and planet as citizens. However, except for Singapore, citizens perceived their national governments and large corporations as lagging in commitment and expected them to assume a larger responsibility in protecting the environment. The majority of citizens who participated in the survey stated that they were willing to accept the stricter environmental rules and change their lifestyle habits. However, the response in favor of lifestyle and habits change is significantly smaller. Respondents are of the view that the solution to climate change should be one of policy, innovation, and technology, keeping personal change and disruption minimal.

A similar finding was mentioned in Kantar's Asia Sustainability Foundational Study 2021. Conducted in Singapore, Malaysia, the Philippines, Thailand, Indonesia, Vietnam, India, Japan, and South Korea, the study found that 58% of Asia consumers are personally affected by environmental issues, including extreme weather events. In addition, 53% of Asian consumers have stopped buying products that harm the environment. However, 63% of Asian consumers do not feel that sustainability is their responsibility, and it is up to organizations such as the governments and corporations to take the lead.

"Consumers want brands to embrace this new mindset and sensibility and help them in their journey. They are looking for social and environmental purpose. And the most active of consumers are influencing others to start thinking this way as well."

— Kantar Public
"What does sustainability mean to Asian consumers,
and what should brands do?"

We sat down with Dr. Casper Durandt, Sustainability Director of Coca-Cola Asia and South Pacific, to discuss how Coca-Cola is approaching the issue of climate change and decarbonization and how he invites his customers and consumers to join this journey.

Dr. Durandt is also a veteran corporate executive. Over the years, Dr. Durandt has founded many successful recycling organizations with a robust business model and value chain that recycles glass to aluminum to plastic. He also serves as an advisor to numerous Producer Responsibility Organizations in Malaysia, Thailand, and Vietnam.

Dr. Durandt received his Ph.D. in Coupled Manufacturing Efficiency Improvement from the University of Stellenbosch in Cape Town, South Africa.

The journey to decarbonization could be long and costly to business. How are you preparing Coca-Cola for this?

Coca-Cola has long been committed to sustainability. Our history of sustainability can be dated back to more than 100 years ago, when we first partnered with The Red Cross in 1917, during World War I. Today and across over 200 countries where we operate, we are investing to address and enhance critical matters concerning environmental and social issues such as water, women, community well-being, sustainable packaging, and climate protection, to name a few.

On climate protection and decarbonization, we constantly assess our carbon footprint, plan, and invest to ensure that our actions align with the ever-growing climate science knowledge while doing our part to maintain global temperature increase to no more than 1.5°C.

As early as the late 1990s, our company embarked on a journey to track energy usage from our production facilities worldwide. In 2013, with the development of our **"drink-in-your-hand"** target, we aimed to reduce the carbon footprint in our products by 25% per liter by 2020, a target which we have achieved in Asia and the South Pacific.

The "drink-in-your-hand" target was Coca-Cola's early attempt to include consumers on the decarbonization journey by raising awareness on the GHG emission of beverage consumption and actions to curb such emission.

I also want to share my thoughts on your question about the cost of decarbonization on business. Decarbonizing our operations and value chain in Asia and the South Pacific will require a change from the "business as usual" mode of operation. While there will be some productivity and cost-saving aspects in switching to energy-saving and renewable electricity, additional investment is required, which we are wholeheartedly committed to. I don't think addressing the climate change issue from the Corporate standpoint can be costless; however, it will become the price of entry for all businesses pursuing growth in the future. Research conducted in Asia suggested that consumers demand products and services that cause less or no harm to the environment. The governments are moving in the same direction. The climate crisis is out there, we know where the game is going, and we want to get ahead. We want to build a healthy business and a healthy planet.

Talking about getting ahead, the "drink-in-your-hand" 25% reduction per liter was undoubtedly a step forward. Is it not quite as progressive as an absolute reduction that is being pursued by other leading companies?

We are evolving our climate goals to align with current climate science. Having achieved our **"drink-in-your-hand"** target in 2020, a new target has been set for 2030. Coca-Cola is currently pursuing the Science Base Target (SBT) to reduce GHG emissions by 25% across the entire value chain by 2030 compared to 2015 emissions.

In Asia and the South Pacific, the SBT will guide and accelerate our climate actions. The SBT differs from the previous "drink-in-your-hand" target. Instead of a relative per liter target, the SBT 25% reduction target is now absolute regardless of our volume growth.

Our climate protection efforts in Asia and the South Pacific will focus on:

- Decarbonizing direct operations by switching to renewable electricity and low carbon energy sources as we are pursuing renewable self-generation capability across Asia.

- Accelerating investments in sustainable packaging, reducing the use of fossil-based material while increasing recycled plastic or rPET use and removing plastics from secondary packaging.

- Investing and switching to 100% energy-efficient and eco-friendly coolers by 2030.

- Reinventing ingredients' supply points and supply chain with more emphasis on Asian sources to reduce the risk of supply chain disruption and lower carbon footprint from logistics and transportation.

We have prioritized recourses dedicated to these climate protection efforts. As a result, they are recognized among key business and commercial priorities with equal importance.

How would you rate consumers' level of awareness, understanding, and willingness to take action concerning climate change in Asia and the South Pacific?

Consumers overall are becoming aware of climate change. People start noticing the changes in nature — the bush fire in Australia, floods in China, and increasing extreme weather events in the Philippines, Vietnam, and Thailand. Our consumers realize that this is a consequence of human activities. Therefore, consumers will be quick to act. There is no doubt that consumers are willing to take action, but coming up with an effective solution to such a complex issue as climate change is no simple task to undertake. Therefore, we believe that our responsibility and role are to offer and promote solutions that consumers can effectively integrate into their means and lifestyle to empower them to become the leaders of climate actions.

One example is our World Without Waste Vision. We are aspired to collect and recycle a bottle or can for every can or bottle that Coca-Cola sells by 2030. Through this Vision, we engage bottlers and invest in Recycling plants in Indonesia (Amendina) and the Philippines (PetValue) to kickstart the value chain and inspire consumers to participate in recycling activities, and stop valuable resources from ending up in our oceans or landfills.

In 2020, the Coca-Cola Beverages Philippines signed a joint-venture agreement with Indorama Ventures to establish PETValue, the Philippines' largest, state-of-the-art bottle-to-bottle recycling facility. PETValue will deploy cutting-edge technologies and industry best practices to recycle plastic bottles made from PET material. The facility could process up to 30,000 MT per year or almost 2 billion pieces of plastic bottles, with an output of 16,000 MT per year of recycled PET resin.

Furthermore, Coca-Cola is also an active member of the Producers Responsibility Organizations (PROs) in Thailand, Vietnam, Indonesia, Malaysia, and Australia. The PROs are organizations founded by like-minded consumer goods companies sharing the common aspiration to enhance the recycling industry's commercial viability and sustainability across Asia and the South Pacific. As a result, a sustainable recycling industry will rely less on virgin plastic from fossil fuels and more on recycled plastic.

These platforms will help accelerate recycling activities and, in return, link consumers to recycling and the reduction of carbon emissions. As a result, we hope to inspire Asia and the South Pacific consumers to help us achieve the climate protection commitment by 2030.

Consumers use billions of pieces of plastic packaging every day. How much can recycled plastic packaging help reduce GHG?

On issues such as climate change and plastic waste, while not explicitly contained to geographies, designing effective solutions will require a clear understanding of the local nuance.

In developing Southeast Asian economies such as Thailand, Indonesia, Malaysia, and Vietnam, recycling plastic packaging represents a significant opportunity for Coca-Cola to reduce GHG emissions further. However, the system in which the post-consumer packaging is collected, sorted, processed, and recycled can vary between countries. Depending on these variations, GHG reduction opportunities are also varied. Given the current mechanical recycling infrastructure and technology available in Southeast Asia, we anticipated the size of the GHG emissions reduction opportunity from switching to recycled plastic packaging to be around 40% compared to the equivalent use of virgin plastic packaging.

Furthermore, we are working with our bottlers and customers to further reduce GHG emissions across Asia and the South Pacific by replacing 100% of our legacy Hydrofluorocarbons coolers (HFC) with Hydro Carbon (HC) coolers by 2030. The new Hydro Carbon refrigerant will be based on Propane and Iso Butane, which are much more environmentally friendly than Hydrofluorocarbons. The new refrigerant is less harmful than conventional refrigerants so far as the GHG impact goes. Together with new technology advancements in the compressor, insulation, and smart coolers, energy efficiency improvements of up to 30% are easily achieved. With this new cooler technology, our consumers and customers can rest assured that they can continue to enjoy ice cold beverages from Coca-Cola without adding more GHG into the atmosphere.

Author's Note

Consumers' expectations are evolving at an accelerated speed fueled by, among other things, the unfolding impact of climate change. The ability to consistently meet the ever-changing expectations will allow companies and brands to be ever-relevant as they pursue sustainable growth. Because climate change is everyone's business, it may not be enough for companies to pursue carbon reduction and climate protection measures within the four walls of business operation. Instead, companies and brands can scale the climate actions impact by inviting consumers to rethink their habits and introducing a new user experience that amplifies the sense of purpose and empowerment.

Endnotes

1. Food and Agriculture Organization, "Leveraging innovation and technology for food and agriculture in Asia and the Pacific", Bangkok, 2020.
2. Ibid.
3. McKinsey Global Institute, "Climate risk and response in Asia", November 2020.
4. United Nations, "Water for life decade: Asia and the Pacific", https://www.un.org/waterforlifedecade/asia.shtml.
5. United Nations Environment Programme, "Water — resource efficiency in Asia and the Pacific", Bangkok, 2011.
6. Rabobank, "Asia Pacific: Agricultural perspectives", 18 February 2016.
7. United Nations, "Goal 2: Zero hunger", https://www.un.org/sustainable development/hunger/.
8. United Nations, "Population 2030: Demographic challenges and opportunities for sustainable development planning", New York, July 2015.
9. Brookings Institute, "The unprecedented expansion of the global middle class", Washington DC, February 2017.
10. McKinsey & Company, "No ordinary disruption: The forces reshaping Asia", September 2015.
11. PwC, Rabobank, Temasek, "The Asia food challenge: Understanding the new Asian consumer", September 2021.
12. Organisation for Economic Co-operation and Development, and Food and Agriculture Organization, "OECD FAO agricultural outlook 2021–2030", July 2021.
13. Research Dive, "The Southeast Asia meat product market: 2019–2026", November 2021.
14. Asian Development Bank, "Asian water development outlook 2016", Manila, August 2016.
15. Christopher Bren d'Amour, Femke Reitsma, Giovanni Baiocchi, *et al.*, "Future urban land expansion and implications for global croplands", *PNAS*, **114** (2006) 8939–8944.
16. McKinsey Global Institute, "The Archipelago economy: Unleashing Indonesia's potential", September 2012.
17. Organisation for Economic Co-operation and Development, "Agricultural policy monitoring and evaluation 2021", June 2021.
18. Asian Development Bank, "Asian water development outlook 2021 update", Manila, September 2021.

19. Reuters, "Nitrogen emissions from rising fertiliser use threaten climate goals", 7 October 2020.
20. Food and Agriculture Organization, "Emissions due to agriculture: Global, regional and country trends 2000–2018", Rome, 2020.
21. Reuters, "India allocates extra $8.71 billion in fertilizer subsidy", 12 November 2020.
22. Farm Animal Investment Risk and Return, "Food giant pledges undermined by "plodding" meat and dairy industry on Covid 19 and climate", 11 November 2020.
23. Ibid.
24. Global Times, "Torrential Henan rains spark concern over grain production", 26 July 2021.
25. Reuters, "Heavy rain in India triggers floods, landslides; at least 125 dead", 24 July 2021.
26. Aljazeera, "Philippines evacuates thousands as monsoon rains flood Manila", 24 July 2021.
27. The Third Pole, "Central Asian drought highlights water vulnerability", 12 July 2021.
28. The Korea Post, "We are committed to advancing 'digital agriculture' to give benefit to our farmers", 16 November 2020.
29. EU Japan Centre for Industrial Cooperation, "Smart farming technology in Japan and opportunities for EU companies", Tokyo, January 2021.
30. GlobeNewswire, "Agriculture drone market worth $5.19 billion by 2025", 9 June 2020.
31. Pinduodo, "Agricultural drone sales surge in China as farmers embrace technology to raise productivity", 14 April 2021.
32. AgriBusinessGlobal, "Outlook 2020: Is this the year for precision agriculture?", 14 January 2020.
33. Abhishek Rajan, Kuhelika Ghosh, and Ananya Shah, "Carbon footprint of India's groundwater irrigation", *Carbon Management*, **11** (3) 265–280.
34. Sharada Balasubramanian, "Solar pumps and drip irrigation help Indian farmers save water and energy", *Rural21*, 15 June 2015.
35. Environmental Defense Fund, "Global risk assessment of high nitrous oxide emissions from rice production", New York, September 2018.
36. Sustainable Rice Platform, "Sustainable rice platform", https://www.sustainablerice.org/.
37. Olam Group, "Olam urges international brands to adopt new rice eco label to allow consumers to support farmers and promote sustainable rice production", 18 September 2020.

38. Walmart Inc., "Walmart sets goal to become a regenerative company", 21 September 2020.
39. The Nature Conservancy, "Syngenta: Collaboration for sustainable agriculture", https://www.nature.org/en-us/about-us/who-we-are/how-we-work/working-with-companies/transforming-business-practices/syngenta/.
40. Cision PR Newswire, "Harmless harvest and its partners launch the first coconut regenerative agriculture initiative", 9 December 2020.
41. Alcimed, "3 ways Asia is cutting down food loss and food waste along the value chain", 23 February 2021.
42. Ibid.
43. International Energy Agency, "India Cooling Action Plan (ICAP)", https://www.iea.org/policies/7455-india-cooling-action-plan-icap.
44. Solar E. Energy, "Solar energy for cold stores", https://www.solar-e-technology-bd.com/solar-energy-cold-storage.
45. S K Cold Chain, "The 3PL of the Future", https://skcoldchain.com.my/.
46. Food Packaging Forum, "Asian governments aim to expand recycled content in plastic FCMs", 24 November 2020.
47. The Straits Times, "Longer shelf life for food? It's all a part of this package", 3 March 2017.
48. Green Queen, "8 Asian food tech start ups innovating the future of food", 30 November 2020.
49. Winnow Solutions, "How governments around the world are encouraging food waste initiatives", 23 August 2019.
50. The Guardian, "The smart tech startup helping restaurants cut food waste by 50%", 27 May 2016.
51. Business & Sustainable Development Commission, Temasek, AlphaBeta, "Better business, better world: Sustainable business opportunities in Asia", June 2017.
52. Microsoft, "Indoor vertical farming in Asia and beyond: Digging deep in data", 27 February 2018.
53. Quartz Media Inc., "The indoor urban farm startup that's undercutting importers by 30%", 31 March 2021.
54. The Straits Times, "Next gen farming concepts on show at exhibition", 7 September 2017.
55. SPREAD, https://spread.co.jp/en/.
56. MIRAI, "Welcome to 'Mirai'", https://miraigroup.jp/en/.
57. Noonthy Greenhouse, "Company introduction", http://www.noontygreenhouse.com/.

58. Business Wire, "Asia Pacific vertical farming produce market (2020 to 2026)", *Business Wire*, 4 January 2021.
59. Nikkei Asian Review, "Japan battles food waste in production with AI and weather data", 29 September 2019.
60. Time, "This startup founder's AI powered garbage cans are helping to reduce food waste — and improve bottom lines", 13 October 2021.
61. WOWTRACE, "WOWTRACE successfully applies blockchain to the traceability of Vietnamese chocolate", 8 December 2019.
62. Food and Agriculture Organization, "Pig farmers in Papua New Guinea capitalize on blockchain technology", https://www.fao.org/in-action/pig-farmers-in-papua-new-guinea/en/.
63. World Wide Fund for Nature, New Zealand, "New blockchain project has potential to revolutionise seafood industry", https://www.wwf.org.nz/what_we_do/marine/blockchain_tuna_project/.
64. HSBC, "Climate asset management", https://www.assetmanagement.hsbc.co.uk/en/institutional-investor/investment-expertise/sustainable-investments/climateassetmanagement.
65. Rabobank, "Dutch government and Rabobank announce anchor investments in AGRI3 fund", 23 January 2020.
66. Bloomberg Green, "Lombard Odier launches $400 million natural capital equity fund", 16 November 2020.
67. GrowAsia, "Agritech business model showcase: Peer to peer lending", https://www.growasia.org/peer-to-peer-lending.
68. Joc Devanesan, "How agritech solutions are shaping Myanmar's digital economy", *Techwire Asia*, 24 June 2020.
69. Blended Finance Taskforce, "Ergos communal warehousing", https://www.blendedfinance.earth/shared-services/2020/11/12/ergos-communal-warehousing.
70. Business Mirror, "CARD Pioneer's crop insurance: More affordable than a pack of cigarettes", 24 October 2016.
71. United Nations Development Programme, "New insurance product to aid fight against climate change in the Pacific", 2 September 2021.
72. S&P Global Platts, "Soil carbon credits: The realities on the ground", 18 August 2021.
73. Sumitomo Corporation, "Collaboration agreement for agricultural soil carbon storage", 22 April 2021.
74. McKinsey & Company, "Agriculture's connected future: How technology can yield new growth", 9 October 2020.

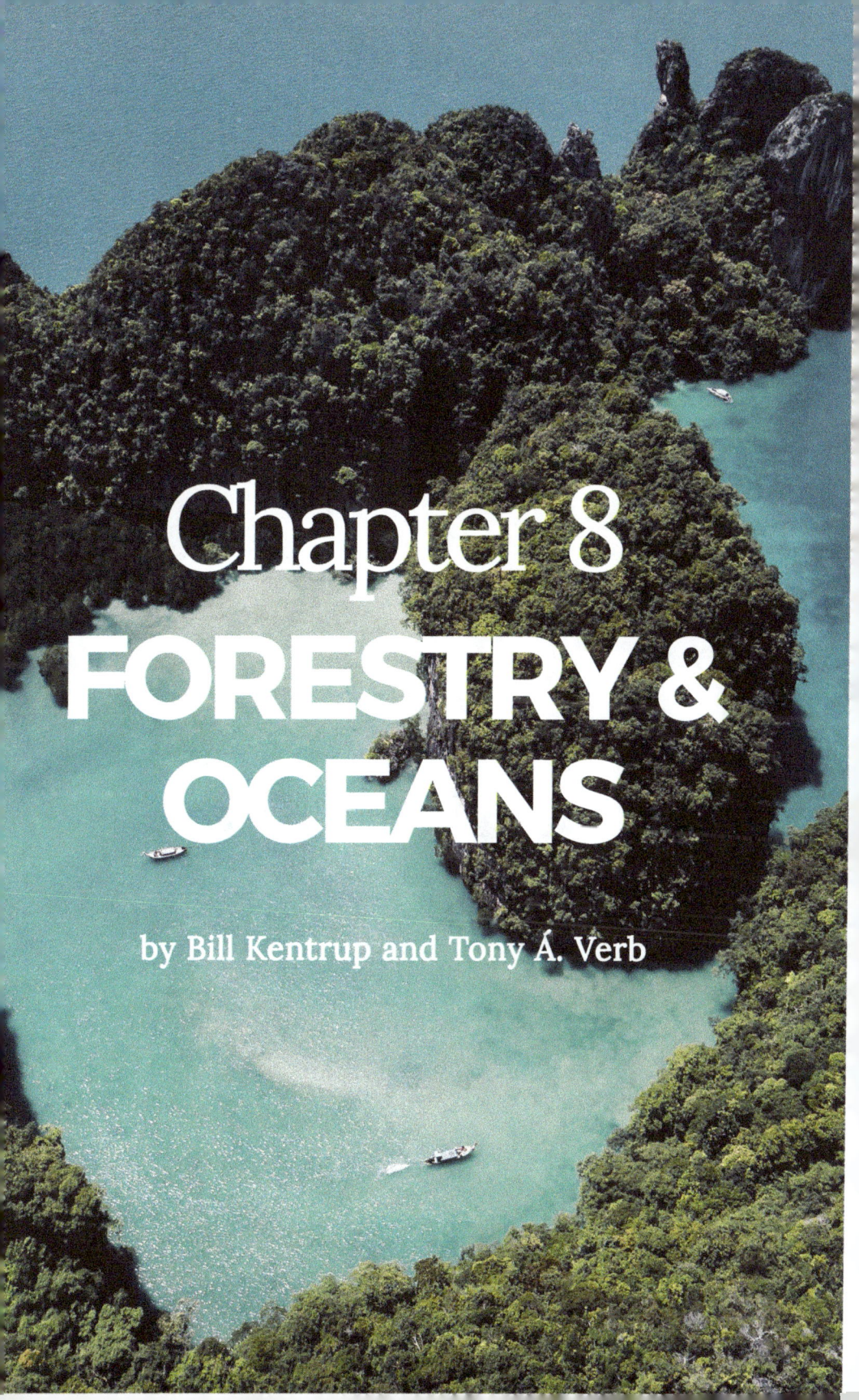

Chapter 8
FORESTRY & OCEANS

by Bill Kentrup and Tony Á. Verb

ANALYSIS

by Bill Kentrup

Not very far from us, but often far from our thoughts, is the constant vastness of outer space. Much closer, but in some ways just as confounding, are the many thousands of kilometers of intensely heated metals, minerals, and ores beneath the Earth's crust. Sitting between these two indomitable and vital domains is a thin collage of forests, grasslands, oceans, rivers, wetlands, breezes, and currents — all mottled with life; in other words, our ecosystem. Whether we go about life as farmers, fishers, shopkeepers, bankers, industrialists, politicians, coders, gamers, or artists, we all depend on the health of our forests and oceans, not just for resources but as perhaps the most significant buffers between our fragile lives and runaway shifts in climate.

Just as the overview effect inspires an appreciation for the fragility of our existence and a sense of helplessness, the scale of our ecosystem, together with the complexity of cycles that drive CO_2 between our atmosphere and forests and oceans, can inspire a similar sense of helplessness for many. But the scale and beauty of our nature and the ingenuity required to do anything sensible and impactful, continues to be an inspiration to many, including the many scientists, engineers, inventors, and investors that have taken on the challenge of supporting our forests and oceans to combat climate change.

Some Forest and Ocean Carbon Basics

CO_2 is constantly exchanged between our forests, oceans, and the atmosphere — CO_2 in the air is removed by plants through photosynthesis via trees and grasses in forests and phytoplankton in the oceans. As plants die and decompose, respiration releases CO_2 back to the atmosphere — directly by forests and through the physics and chemistry of surface and deep-water column that makes up our oceans.

With the exception of carbon caught up in the Earth's crust and mantle, our forests and oceans represent the vast majority of the Earth's carbon stocks. The amount of CO_2 trapped in the waters is nearly 40,000 gigatons of CO_2 on average. Around 4,000 gigatons are held by forests, vegetation, and under permafrost. Putting that into context, the burning of fossil fuels contributes just over 6 gigatons of CO2 per annum. Whilst oceans and forests have the capacity to store carbon at a level that eclipses the annual CO2 released by the burning of fossil fuels, the absorption rate cannot keep up with the emissions rate.

In terms of carbon flux,[1] it is estimated that close to 80 billion metric tons (MT) of carbon pass into and out of the world's oceans each year, with the balance being slightly in favor of carbon entering the ocean, making the oceans a net carbon "sink" by approximately 2 billion MT. Similarly, our terrestrial ecosystem absorbs carbon at a rate of 2–3 billion MT per annum more quickly than carbon is released, with an annual exchange in the range of 120 billion MT. However, the oceans and terrestrial landscape are not always net carbon sinks, and the conditions necessary for them to become net carbon sources are taking shape. With cumulative flux into and out of the oceans and atmosphere on the order of 200 billion MT per year, any small shift towards becoming a net carbon source would seem to imply a few tens of billions of MT per year. But, taking into consideration that the total carbon stocks of oceans and forests are in the tens of trillions (and that in our atmosphere is less than one trillion), forests and oceans are important in the global carbon cycle (see Figure 8.1) and in our strategy to combat climate change.

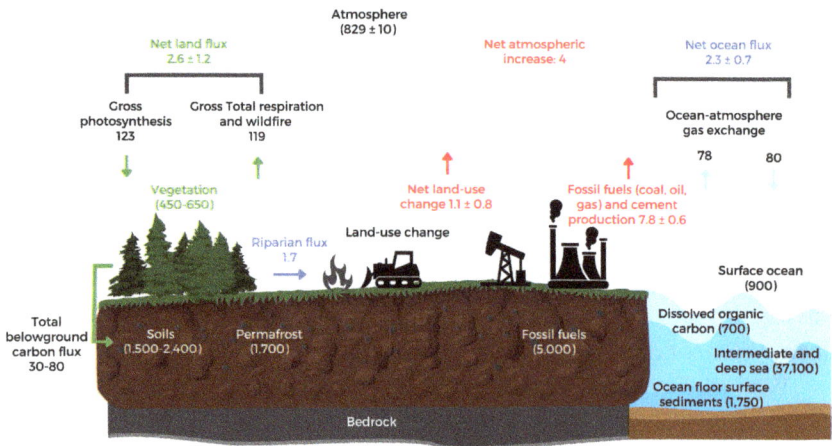

Figure 8.1.   The global carbon cycle.[2]

The residence time[3] of carbon (i.e., the time carbon enters a particular system to the time it once again exits that system) is on the order of decades (i.e., forests), centuries (i.e., soil) and millennia (i.e., oceans), although the residence time for oceans varies — hours on the surface, 300 years at mid-ocean depths, ~1,000–2,000 years for deep oceans (see Figure 8.2). This gives hope to climate scientists that the carbon sequestered into our forests and oceans can potentially be kept out of the atmosphere for a long time. Sadly, however, carbon emissions sequestered through forestry have decreased significantly in the past 20 years due to deforestation from logging, plantations, agriculture, and construction, as well as an increasing number and scale of fires due to global warming. Striking a healthier balance between human activity and our forests and oceans is not just important for carbon's sake, it is also vital for biodiversity, natural resources, food, tourism, medicine, and the livelihoods of hundreds of millions.

Both forest and ocean ecosystems help to absorb some of the CO2 that we have been emitting, but the rate of new CO2 emissions per year outpaces the rate of absorption. A number of factors are currently tipping the balance in the wrong direction.

As global average temperatures warm, the ocean warms — thankfully, very slowly — and the ocean's ability to retain dissolved gases of any

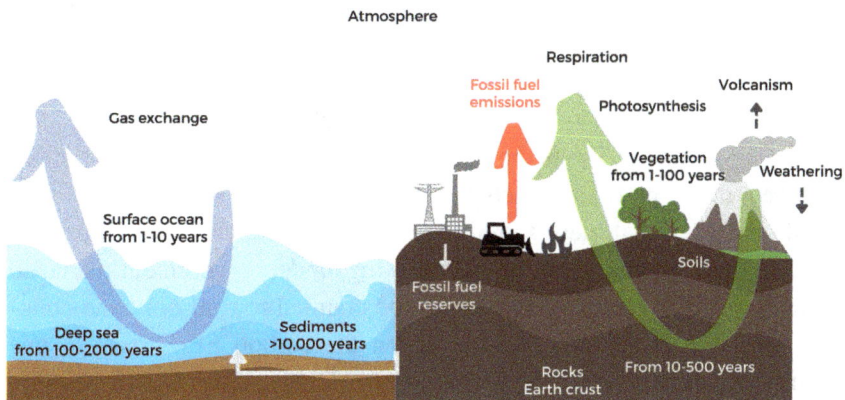

Figure 8.2. A simplified schematic of the global carbon cycle showing the typical residence time scales for carbon transfers through the major reservoirs.[4]

kind, including CO_2, is inversely related to temperature. Nothing is surprising about this because when liquids warm, they release gas, and some of that gas — in the ocean's case — is CO_2. Also, as forests, grasslands, soil, and similar terrestrial ecosystems are cleared or degraded, they, too, have a reduced ability to absorb CO_2. The latter is happening on a global scale and significantly so in Asia-Pacific.

What all this means, from the perspective of the business of climate change mitigation, is that any proposition that has the potential to tip the balance in favor of more carbon storage is of great potential value.

Where Forests, Oceans and Carbon Finance Meet

When we refer to "carbon finance", we mean capital that is invested with the explicit goal of reducing emissions. Carbon finance can take the form of equity, debt, or hybrid instruments where the capital provider is entitled to a portion of the claims relating to the emission reductions achieved. Carbon finance can also take the form of the sale and purchase of carbon offsets, carbon credits, and other carbon-related environmental financial products or "carbon EFPs". These may be structured as payment on delivery, advanced payment for future delivery, options for the right to a future purchase of carbon EFPs, and so on. Essentially, carbon finance takes many of the forms that we see in traditional finance and trading, but where a core component of value is being created and transferred, is the ability of the financier to claim the impact of reduced emissions — represented by a carbon EFP.

Where a party acquires the rights to a carbon EFP, that party — and only that party — becomes the only party that can validly claim that it has achieved those particular metric tons of CO_2 emission reductions. It should be noted that these same metric tons of CO_2 emission reductions may, for the purpose of carbon reporting under government-managed frameworks like the Paris Agreement, be counted toward only one country's emission reductions — even when the carbon EFP arises in one country and is transferred to another. For example, in a case where party A in country X sells a carbon EFP to party B in country Y, it is clear that party B has the right to claim that emission reduction as long as it meets the emission reduction criteria party B is reporting under. But, whether it is country X or country Y who may report that emission reduction as falling within their "nationally determined contribution" or country target set up under the Paris Agreement is the subject of ongoing discussion and negotiation amongst the countries that have signed on as parties to the Paris Agreement.

By definition, carbon EFPs are products that embody the evidence that a certain amount of greenhouse gas — typically denominated as one metric ton of CO2 equivalent or tCO2e — was avoided, reduced, or removed from the atmosphere, and that it was achieved at a particular location, by a particular means, at a particular point in time. With a metric ton of CO2 measured, verified, and reported in accordance with a market standard, an offset can be created and purchased by governments, companies, or consumers, thereby connecting the dots between the financing of CO2-friendly activities and the rights to claim the positive impact against one's carbon footprint. Financing carbon offsets is one of a number of important tools used to achieve decarbonization and, more broadly, sustainability goals.

Subject to meeting a raft of criteria that is necessarily adjusted over time, particularly to account for additionality,[5] assets that give rise to carbon EFPs may cover hundreds of different types. Some more familiar types include renewable power, heating and cooling, energy efficiency, methane avoidance and/or capture from landfills, organic effluent, coal mines, and oil and gas fields, and the capture and destruction of potent industrial greenhouse gases. Some less popular or even esoteric initiatives that have been tried (with varying successes) include: switching light bulbs and shower heads to more efficient devices, switching from coal-fired stoves — a key contributor to respiratory illness in some countries, and ventilated biomass stoves.

Across the realm of forests and oceans, there is a wide range of land use and forestry practices that could potentially result in measurable and verifiable (i.e., real) carbon offsets. Together, we refer to these as "nature-based" (NB) or "natural capital" carbon. On land we have avoided deforestation, afforestation, reforestation, and grassland and wetlands rehabilitation. Within each of these, there are many approaches and methodologies that may give rise to NB carbon EFPs. In the case for oceans, there are strategies that would mechanically pump large volumes of CO2 down to the ocean depths, to those that would increase the rate of primary productivity, or plant matter growth, sucking carbon during photosynthesis from surface water. In the latter example, the same plant matter, together with the CO_2 it has absorbed, would sink to the ocean depths through natural or facilitated transport. Such strategies would result in large volumes of CO_2 that would not be seen back in the atmosphere for hundreds to over a thousand years — or at least that is the scientific possibility and the goal.

Before exploring the past, present, and future of forests and oceans in global carbon markets, here is a short note on why it is beneficial for EFPs to exist amongst the tools being used to manage climate change: Greenhouse gases, regardless of where they are emitted, have mostly a global impact by increasing the total heat trapping capability of our atmosphere. Related to this, the positive impact from a climate change perspective that is realized through the reduction of one metric ton of CO_2 from the atmosphere is similar across the globe, whether it happens in the European Union (EU), Asia, the Americas, or Africa. However, the cost required to reduce one metric ton of CO_2 can vary from less than USD 1 to as much as hundreds of US dollars, and each country and industry will have a certain commercial capability to contribute toward reducing climate change. To put it another way: the greater the cost, the more the resistance, and the slower the action.

And so, allowing countries and corporates to collaborate on reducing emissions and deploying capital into projects that reduce more emissions per USD 1 spent theoretically drives more or quicker emission reductions. This approach and practice, largely through technology transfers, can set the groundwork for long-term low-carbon economies where they might otherwise not exist.

This is cap and trade — this is the carbon market.

PAST

To orient ourselves, in the case of forests, much of the carbon storage capacity sits in the tropics, in particular, South America, Central Africa, and Southeast Asia. To a lesser degree, forests in northern latitudes store carbon as they grow and develop, though the rate at which they grow is slower, and the amount of carbon they store is less than in the tropics. In both the tropics and temperate areas, forests sequester and store carbon as they grow — sequestration slows but continues as they mature. Grasslands store less carbon per unit area relative to forests, but they cover vast areas; hence, the potential to store carbon in well-managed grasslands is meaningful.

Some of the earliest advocacy for NB carbon EFPs took place over 40 years ago, and early demand for forest carbon offsets started to take shape in the late 1990s, with the commitments and carbon trading mechanisms contemplated under the Kyoto Protocol. It had been the Kyoto markets, starting up in earnest around 2005, which propelled the demand for a wide variety of carbon EFPs, including forestry carbon EFPs. However, voluntary markets preceded the Kyoto markets — these continuously move forward, sometimes running, sometimes crawling. They do so whether or not global governments reach a consensus to enter into agreements like the Kyoto Protocol and the Paris Agreement, which are both managed by the United Nations Framework Convention on Climate Change (UNFCCC).

PRESENT

To date, dozens of forest carbon projects have been developed and approved under the UNFCCC Clean Development Mechanism (CDM) and several hundred developed under the major voluntary market standards. However, it should be noted that many regulation-driven compliance carbon markets accept certain carbon EFPs from voluntary carbon standards. Likewise, it is not uncommon for parties meeting voluntary carbon goals to invest in carbon offsets issued under the CDM. As such, the significantly larger volume of projects registered by voluntary standards relative to government-driven standards may understate the volume of demand and project development buoyed by the existence of government-regulated markets.

Taking a closer look at Asian voluntary markets, as of May 2022, there were 67 projects in the Agriculture, Forestry and Other Land Use category registered on Verra,[6] with another 117 in the pipeline. An impressive 135 of these projects are in China. At the same time, Gold Standard[7] had 11 Land-Use Activities and Nature Based Solutions, with only one in Asia, although Gold Standard has been aligning toward greater participation in the "natural capital" segment of the carbon market.

We also see compliance markets for forestry carbon in Asia, including China's national market and three regional markets, Japan's national market and three prefecture-level ones, as well as Korea's Offset Crediting Mechanism. The Korean market has been posting some of the most consistently high carbon prices globally since 2016, at times exceeding the European emissions trading system (ETS) price. Incidentally, due to the allowance of certain types of internal carbon EFPs in the Korean ETS, an arbitrage opportunity run by several traders was to buy and sell Korean-eligible EFPs from the EU market into the Korean market.

The action on the ground within China in relation to land-related carbon, in general, has been positive, with material increases in forest cover and impressive tree planting programs undertaken in recent decades. An average of 600,000 hectares per year of afforestation in Inner Mongolia is being carried out, with current plans to plant close to 100 million mu (6.67 million hectares) per year of afforestation, reforestation, and grassland rehabilitation.

In recent years, national and sub-national carbon markets have increased the representation of forestry carbon projects, and in some markets, forestry carbon EFPs are the most prevalent type of carbon EFP, driven by local co-benefits[8] derived by these projects.

The acceptance of forestry carbon EFPs in international compliance markets is variable, for instance, REDD+[9] EFPs are not eligible in the EU ETS, one of the largest and most consistent compliance markets globally. They are, however, at the time of writing, allowed for use under CORSIA, the framework governing carbon for international air travel. In the voluntary markets, forestry carbon projects have historically represented 1/3 or less of all carbon offsets transacted, except in the past 3–5 years, where that percentage has been around 40%.

Nature-based Carbon — A Landscape of Challenges

Nearly all segments of the carbon market have gone through growing pains, and one might say these growing pains will never be "outgrown" completely and are inherent to carbon finance. NB carbon projects have been characterized by a particularly long and diverse history of challenges.

As with many large scale, high impact, and long-term endeavors, a NB carbon project must have financial backing — regulatory or charitable in origin or purely commercial with risk-adjusted returns — that is sufficient to attract investment. Whilst financial backing is important, the topics discussed in this chapter generally concern projects implemented commercially, supported by carbon finance. Hence, to a reasonable degree, the challenges here are discussed with respect to investability.

For NB carbon projects, given the vast expanses of land and diverse development and operational activities typical of these projects, the list of "what could go wrong?" is quite long. Here are some examples: Has the project conducted the appropriate due diligence and feasibility work? Has it obtained the correct permits, approvals, rights-of-way, and impact assessments? Are the regulatory bodies of concern national, provincial or local — where should one go to find out what is actually required? Has the project put in place an adequate financing plan for development, implementation, and operations? Are experienced contractors identified and under contract with in-market terms for underperformance? Are appropriate training and incentive mechanisms in place for staff and other local stakeholders? Are there risks of illegal logging, settlement, fire risk, disease, and insects, and what proactive or reactive measures can be taken to mitigate these? Has the project budgeted and structured for normal operations and for reasonably anticipated downside scenarios? The list goes on.

These may all come across as normal and, in some ways, mundane business development and operational issues. However, global carbon pricing and other carbon-related incentives have not been consistently high enough for long enough to support a flourishing forest and ocean carbon industry. Hence, the presence of long-standing, specialized and dedicated platforms managing these issues are few and far between. This market gap is at times fully appreciated only after significant time, good-will, capital, and climate claims have expired.

With carbon market volatility, it has been hard for forest projects to be supported by purely carbon financed business models. There are some examples across Indonesia, Papua New Guinea, Malaysia, and Philippines of integrated agriculture + forestry or peatlands + forestry projects. These models come with more diversified revenue streams and/or higher rates of carbon EFP productions due to methane avoidance, which means they can potentially be sustainable at relatively low carbon prices.

For those projects successfully operationalized, not only must a NB carbon project demonstrate that carbon has been reduced, removed, or avoided, but that reality must be seen as permanent on a time scale of decades, if not centuries. If a forest is degraded in terms of the amount of carbon retained, or if the carbon sequestered in the ocean is once again released into the atmosphere, this could retroactively impact related carbon EFPs previously issued by the project. What happens to any offset claims made in respect to that project then becomes subject to government or market adjustments, penalties, reputation risk, and so on.

All of the above falls into the category of critical topics that need to be addressed during the development and operation of large-scale forestry carbon projects, and these are just the headline topics, with many more "devils in the detail" showing up over time.

Some of that bedeviling shows up in the monitoring, verification, and issuance of carbon EFPs. Purchasing carbon EFPs to reach corporate sustainability goals are only meaningful if they are measurable and verifiable. But given the technology available in the past, data collection and analysis of carbon-offset projects have historically been a complex, slow, and manual process. It can take years for newly developed and invested forest carbon projects before the first carbon EFP is issued — this allows a period for new projects to reach a level of stability while requiring asset management and monitoring. Once the costly process for verification and issuance of carbon EFPs begins, it can take many months and with some unpredictability in terms of how many final EFPs will be issued.

There have often been severe bottlenecks in terms of qualified experts available to start and complete the process of creating measurable and verifiable carbon offsets. This is partly due to the vast and potentially heterogeneous land or water masses involved (in the case of oceans), coupled with the acute cycles of concentrated demand for carbon offsets seen across the market; carbon markets can be highly seasonal.

Apart from verifying that emission reductions have occurred within a particular project boundary, it is also important to be certain that "leakage" has not occurred, i.e., a scenario where the carbon sequestered by the project is replaced by an increase in carbon emissions elsewhere. It is easy to envision how the activity for reducing carbon across a particular area of land, e.g., reducing illegal logging, could inadvertently incentivize the use of other neighboring areas of land in less carbon-friendly ways. Trying to track leakage outside the project boundary expands the scale and type of monitoring and verification that must occur — a daunting prospect if done manually.

NB carbon offset projects can be rewarding investments when managed well. But inefficiencies and bottlenecks such as those above can result in financially strained projects. Given the competing uses for land areas, the sooner these bottlenecks are removed, the sooner the financial incentives for carbon-friendly forestry projects improve. While models for ocean carbon development are more nascent, they will surely face similar bottlenecks without the necessary solutions discussed further below.

Finally, in terms of challenges, we must mention carbon accounting. It will not surprise anyone who has been through an audit of any kind that accounting errors, or in some cases willful misrepresentations of carbon EFPs, occasionally occur. Unless each carbon EFP is properly accounted for in terms of which government, corporate, or individual has a right to claim it, forest and ocean carbon EFPs, just like any other carbon EFP, run the risk of being counted more than once.

Given the need for coordinated carbon-reduction efforts at the global, regional, and local levels, and the need for carbon emissions and reductions to be reported as accurately as possible, e.g., under legal obligation by all signatory governments of the Paris Agreement, it is absolutely vital that the data that underpins each carbon EFP is detailed, accurate, and accessible.

Many sensible and not-so-sensible carbon tech solutions are being developed and deployed, with varying degrees of welcome and resistance from the carbon finance industry. Despite the carbon market orientation around a common and important endeavor to keep our world habitable, change in this industry can suffer inertia brought about by the accumulated mass and interests of incumbents. One might argue this is understandable, given the vast field of new entrants — with widely ranging levels of competence — to the market, trying to shake things up at every level.

An example of how some of the gaps discussed above have manifested in the demise of a potentially solid forest carbon project was outlined in ProPublica in 2019.[10,11] In short, the basic lay of the land was:

- A rather large-scale forest protection project in Cambodia initiated in the early years of the Kyoto period was seemingly well planned and backed by the local government and influential capital.

- However, cross-border military clashes, sanctioned logging, village developments, illegal farming, and the presence of land mines all began to negatively and materially impact the smooth implementation of the project.

- For the regions covered under this particular forest protection project, satellite analysis commissioned by ProPublica revealed a significant reduction in forest coverage since its inception, which was precisely the opposite of the hoped-for, planned, and funded outcome.

It is worth noting that some diminishment of the forest is not unusual and may even be included in the project scenario, and a successfully executed project may only slow the rate of forest loss rather than stall or reverse it.[12] Furthermore, the implementation of this particular forest protection project was not described as an instrument of forest degradation, but it was, sadly, unable to preserve the covered forest area.

Whilst some of the problems described in the above example may be somewhat dramatic, far more mundane administrative or operational gaps have historically left many otherwise great forest carbon projects stranded as an unmet or underperforming opportunity.

Back to Oceans

As compared to forests and grasslands, it is harder to draw a line between human activity and carbon sequestration in the oceans. While ocean-related carbon EFPs have been contemplated since the 1990s, active development of businesses oriented around generating carbon EFPs from the oceans took place later. In order to impact ocean carbon, water temperatures need to cool or ocean-directed carbon needs to sink.

Our ability to lower ocean temperatures to enhance CO_2 sequestration from the atmosphere to the oceans is limited, given the volume of water in the world's oceans. And if we could, this could drive other changes that may be undesirable and unpredictable. For example, water temperature is one of the key drivers of oceanic circulation, which involves a conveyor of water currents flowing and water masses rising and sinking at every depth, in every ocean, all the time. If we were to find a way to reduce global average water temperatures, the effects on the oceanic conveyor belt, as well as the biology and weather that depends on it, would certainly be impacted — potentially negatively, potentially positive, or both, depending on the organism.

The current slow rise in oceanic temperatures can (and do) have similarly far-reaching effects. The nature and extent of some of those effects are indeed unpredictable, but the release of gases mentioned earlier as liquids' temperatures rise is perfectly predictable.

In terms of getting CO_2 to sink, models for pumping seawater that is supersaturated with dissolved CO_2 to great depths have been contemplated but not tried at scale.

As for the conversion of dissolved CO_2 to particulate CO_2 — packaging CO_2 into phytoplankton, seaweed or seagrasses, and subsequent sinking — one needs to have the right nutrients for the right plant cells in the right place and at the right time. Humans play a role in affecting nutrient availability along the coasts, with agricultural, sewage, and wastewater treatment regularly feeding "blooms" of phytoplankton. However, considering that the vast majority of coastal plankton is destined to sink in relatively shallow areas, with regular mixing between coastal depths and surface water, the CO_2 sequestered under this model is only taken out of the system on the timescale of hours to perhaps months. And as ocean temperatures have slowly increased, in addition to the release of dissolved CO_2 from the ocean to the atmosphere, research has shown that ocean water temperature increases seen since the late 1980s have led to a reduction of around 1.5% of CO_2 in particles to ocean depths.[13]

It is really the vast expanse and depth of the open oceans that can do the job of long-term removal of carbon via plankton and other vegetation-driven carbon sequestration. Up to now, humans' role in driving or limiting open ocean phytoplankton growth has been very limited. That's not to say humans have not contemplated their role or even tried their hand at "fertilizing" the oceans for the purpose of combating climate change.

Perhaps some of you reading this text are familiar with the often-quoted remark by oceanographer John Martin at the Woods Hole Oceanographic Institute: "Give me half a tanker of iron and I'll give you the next ice age."[14] It is a grand and potentially controversial remark, but underpinned by a hypothesis validated by many oceanographic studies that much of the open ocean is iron-limited in terms of phytoplankton growth. This means that by adding some iron and a certain volume of phytoplankton, we could get their photosynthetic machinery running to draw down significant concentrations of carbon. Then the carbon, along with the dying plankton itself, would sink to the ocean depths.

In practice, a developer called "Climos" had a go at stimulating phytoplankton blooms during the Kyoto markets. This initiative may well have been charting the right course toward cooler times (and hopefully not the next ice age), but overall, the carbon market price did not get high and long enough to allow Climos to get to the ocean depths and operational scale required.

FUTURE

As discussed above, while initiatives to generate EFPs from forests have been undertaken for about 25 years (about 15 years for oceans), creating and using NB carbon offsets have always been challenging within global carbon markets. However, as more companies are making net-zero commitments, the appetite for forest carbon EFPs and other NB EFPs is growing quickly, at least in the voluntary (non-regulatory) carbon market. The Taskforce for Scaling Voluntary Carbon Markets anticipates EFPs from forestry to contribute between 65 and 85% of the carbon EFP supply to the voluntary markets between now and 2030. While ocean sequestration EFPs have historically been nearly non-existent, EFPs from wetlands, marshes, mangroves, bays, and other coastal environments are on the rise. Initiatives like the Seascape Carbon Initiative, launched in August 2021, hints at further activity.

How Technology Can Help Forest and Ocean Carbon Project Development

Technological innovation cannot and will not be able to solve all of the challenges faced by NB carbon projects. However, technology is rapidly becoming part of the solution and is starting to bring about meaningful improvements to offset projects and markets.

The following are a few tech-based improvements that are available now and being deployed in some regions across the world.

Remote Sensors, Satellite Observation, and Artificial Intelligence

One avenue being explored and deployed for certain forest and ocean management applications is the use of remote sensing — both through satellite observation and through robots, both ground-mounted and aerial. Such remote sensing, together with specialized analytics, have long brought vast improvement to the agriculture and fishing sectors. It has aided in the protection of vast forests and has also provided invaluable guidance to ocean scientists in their research.

In the case of forests, such tools have dealt with the risky build-up of "forest fuel", early detection of fires, disease, and illegal land use, and have aided in the responses to these risks. It can even help detect the occurrence of the "leakage" mentioned earlier.

Not only can such technologies help reduce the cost and risk for areas managed for agriculture, recreation, or biodiversity, but satellite and robot-obtained tree canopy data, together with soil sensor data, is already being used to measure and monitor carbon stored above and below ground. Similarly, remote sensing helps in the measurement of carbon and carbon flux across ocean depths.

From an environmental, financial product perspective (i.e., the business of financing through environmental offsets), the same improved predictability and reduced operating, financing, and insurance costs have been realized in the agriculture space in particular and is starting to be realized in the carbon offset space. For example, Insight Robotics, based in Hong Kong, has been making use of fixed and mounted robots, mobile drones, satellite data, and data analytics for nearly a decade to help clients manage fire risk, disease risk, pest risk, and illegal logging for forest and agriculture plantations. These improvements have had material success in helping clients in relation to operational costs and productivity. This capability could prove invaluable to the carbon market. More recently, tech-forward companies like Cultivo, Sylvera Carbon, and Regen Network have been exploring and pursuing opportunities to use various sensor and data analytics technology specifically to improve the planning, structuring, and operational reality for forest carbon offset projects.

A tech-forward NB carbon space is one that has been essentially waiting to happen, as soil and water sensors, satellites, and fixed and mobile drones can help provide data-rich products that can verifiably tell a story beyond just carbon.

Blockchain Technology

Historically, it has been difficult to easily and permanently link the data from a carbon offset to the data from verification at a specific location from a specific time, potentially limiting and even hurting the credibility of environmental claims made using such offsets. The need to efficiently and digitally link data to carbon offsets is becoming more acute as carbon accounting for corporations goes digital. The same applies to carbon reporting, relating to country registries, meta registries, and the governing bodies of the Paris Agreement.

A solution that is starting to be leveraged is the use of blockchain technology to record carbon-related data. That is, data collected directly from the devices and sensors described above is permanently and verifiably linked to carbon offsets that have been achieved.

Having data packaged in this way not only creates carbon offsets that carry a digital signature all the way from creation to transfer to retirement, but it also allows offsetting activities to fit neatly into the digital carbon accounting and reporting of the above-mentioned into a growing list of digitized professional services. These include tracking the performance of climate-related financing, reporting between general and limited partners, data for green ratings, and group-wide tracking of decarbonization goals.

A model for how this could be done was demonstrated under Project Genesis, initiated by the Bank of International Settlements and the Hong Kong Monetary Authority. This project demonstrated how blockchains could be utilized by bond issuers to issue tokenized green bond instruments, available to a much wider universe of investors (including retail), and with a data feed from the financed assets providing much-needed transparency to bond-holders about the achievement of intended environmental impact.

Putting carbon-related data on "digital rails" is a way of future-proofing a party's decarbonization activities. In the near-term, it can allow for quicker, cheaper, and higher quality production of carbon offsets and better-structured financing, insurance, and professional services — all absolutely critical to strive for given the urgency with which we must combat climate change.

Even in the People's Republic of China, where the role of blockchain, and in particular, cryptocurrencies, has had a mixed welcome, State-owned entities are exploring the use of blockchain to enable efficiency in environmental product markets. A pilot of such a model in the renewable power space, albeit not yet in the NB carbon segment, was made public in 2021 by China's State Grid and Southern Power Grid. The project was acknowledged by one of China's most senior administrations under the State Council, the National Development and Reform Commission.

What Might the Future of Forestry and Ocean Carbon Finance Look Like?

The exact features of what a tech-enabled forest and ocean carbon future could look like will depend on the direction of environmental markets in general, including policy, project development, and financing. However, taking stock of trends in the space, this future could likely entail increasingly data-rich carbon instruments that are verifiable and traceable in terms of their impact, who financed that impact, and who claimed against it. In the not-too-distant future, we may see large volumes of carbon claims being transferred in the context of tokenized green bonds, tokenized carbon options and futures contracts, and other long-dated carbon finance instruments, and less so under carbon spot contracts. It is worth noting that the shift from projects selling credits once issued under spot contracts toward long-term forward contracts covering several years will enable carbon market activity to grow at scale, regardless of the tech trends described in this chapter.

We may also see projects take advantage of the combined effect of the large volume of plankton, seaweed, and grasses in coastal areas. This will require not only advancement in technology and engineering but also in planning and coordination within and between countries. The ocean is vast, deep, and dynamic, with stakeholders in just about every direction. Might we eventually use digital chemistry or "chemputation"[15] to simply interfere with the heat-trapping properties embodied in the chemical bonds of the greenhouse gases? Whether or not we get as far as "chemputation" for carbon any time soon, there is plenty of value to be unlocked and positive impact to be gained through the gradual and smart integration of technology in the context of NB carbon projects and through leveraging the know-how and presences of climate-aligned infrastructure developers and operators.

STARTUP CASE STUDY
Cultivo

by Bill Kentrup and Tony Á. Verb

Whether land use activities are industrial in scale or localized, for agriculture, livestock, natural resources or tourism, or for modern or more traditional purposes, it is often the case that human-led land-use practices result in one or more detrimental effects on the land those practices depend on. Such practices may result over time in the net release of carbon, the degradation of land for sustained human use, and the deterioration of impacted ecosystems.

It is fairly straightforward to identify models for land use that result in short-term gain by running the algorithm of greatest financial output versus lowest cost. It is indeed much harder and more abstract to grasp models and incentives that promote long-term development and growth of communities and that are in balance with environmental and ecological concerns.

The regenerative land use space has historically suffered from the slow development of methodologies that allow developers to bring forward the environmental and ecological benefits from their assets. It is hard to justify financial investments or commitments to change from conventional practices unless a value can be placed on the result.

Our select technology startup, originally from Mexico — but now a Public Benefit Corporation headquartered in California — is addressing this challenge and opportunity. Cultivo, which describes itself as a climate-focused fintech company, uses satellite imaging technology to identify unsustainably managed land for restoration, conservation, and sustainable use. It deploys methodology and practices during the development and operational phase of natural capital assets.[16] Cultivo participates in this space as an advisor and investor, providing technical and commercial planning as well as carbon verification services. It also helps to monetize environmental financial products related to carbon, water, biodiversity, and other real and verified improvements. Managers and farmers of the improved land share these benefits, aligning local stakeholders toward long-term sustainable land use.

With a curated staff versed in environmental science, finance, and technology, together with boots-on-the-ground, Cultivo's mission is to accelerate investment into nature at scale with the specific goal of "regenerating 1% of the planet's total land surface area".[17] As a commercial proposition, the company makes institutional-grade financial products available to institutions and strategic investors that are eager to invest in natural capital. The company collaborates and partners with landowners, NGOs, project developers, and domain experts in its work.[18]

At the time of writing, the company has more than 20 staff across its markets and has raised more than USD 6 million in seed funding at the topco level, whilst setting up local, specialized joint ventures for each project is part of the strategy.[19] With natural capital development activities spanning Australia, South Africa, Mexico, USA, Indonesia, and Madagascar, Cultivo's partners span local land conservation NGOs like Terra Habitus to global platforms like the World Bank's International Finance Corporation, supported by a board that includes the former CEO of the Nature Conservancy Mark Tercek and serial risk-management entrepreneur Gabriel Holschneider.[20] The company has received recognition from organizations like the World Economic Forum and the Unreasonable Impact Initiative for its work.

As Cultivo brings forward its Asia-Pacific strategy, its relevance to the region is in its solution. Regenerative farming and land use has long been advocated, but few incentives and business models have emerged to encourage its proliferation in the region. To support its expansion, Cultivo is always on the look-out for partners aligned with its mission that can help bring forward natural capital assets in the context of the diverse regulatory environments across markets. With more and more public and private stakeholders in Asia looking for new approaches to sustainable land development, and an increasing level of commitment to net zero in the region, the time to amplify action may be upon us.

CORPORATE CASE STUDY
UPC Renewables

by Bill Kentrup

**Synthesizing the need for a cost-effective and reliable way to imple-
ment natural capital projects is the focus of the first case study of our
chapter.** The core business of UPC Renewables is the development, construction, and operations of renewable energy assets, principally wind and
solar. It is a platform with multi-decades of focused renewable development across North America, Europe, the Middle East, and Asia-Pacific.

Given the competitive nature of the renewables market, all developers
who have aspirations of profitability and longevity need to innovate in
nearly all aspects of their business, whether it be operational efficiencies
that result in saved costs, more power output, or at least more predictable
outcomes, sensible diversification (e.g., power storage, rooftop solar,
floating solar, geothermal power), diversified access to better, more
aligned financing, and generally bringing more to the table locally as they
build their operations in remote regions and across vast land areas.

Critical to the success of renewable energy development is a strong
"ground game" (as UPC puts it) in areas such as:

- Administration: An ability to set up companies, secure development
  rights, obtain permits and approvals, train local staff, identify correct
  on-the-ground incentives, conduct environmental impact assessments
  and biodiversity studies, conduct ongoing maintenance (routine and
  reactive), pay taxes, develop partnerships, etc.

- Carbon asset development: As EFPs, including both carbon and
  renewable energy certificates, can be an important revenue stream for
  renewables in developing markets, the important and mundane capabilities of submitting projects for approval under a particular EFP
  standard, maintaining operations in accordance with the chosen standard, keeping documentation in order, having projects conduct verification and certification of EFPs for issuance and sale, etc. This is all

normal course of business for experienced renewable developers like UPC and involves a hard-learned and localized ground game.

- Forestry and biodiversity: Finally, for renewable developers, looking after, documenting, and reporting on the state of the surrounding forests and biodiversity is historically an obligation to a small or large degree and generally sits as a financial and human resource cost for renewable energy developers.

As discussed earlier in this chapter, many of the often-material gaps in the forestry carbon market historically can be summed up as gaps in the "ground game" necessary to successfully design, implement, and consistently operate forest carbon projects. UPC's solution with its partners has been to pioneer a model for integrating natural capital development (e.g., forest carbon, grassland carbon, carbon-reducing agriculture) with renewable development and operations. This model, successfully implemented, allows forest carbon projects to benefit from all of the generic "ground game" capability, physical presence, and innovation UPC has under the hood. This model may be exactly what the forest carbon market needs for success.

To bring this forward, UPC has been working with various technology companies like Cultivo — an analytics specialist for the development of natural capital assets — to identify attractive opportunities to turn the "cost and compliance" work renewables must undertake relating to forests and biodiversity into an opportunity and revenue driver of significant scale.

Leveraging a platform like UPC Renewables to successfully implement natural capital assets has the potential not only to create value for UPC and its partners but also help local communities and governments to reliably realize the promise of natural capital carbon development, as opposed to seeing yet more projects get stranded due to a lack of "ground game", technology, and innovation.

Countries in Asia where this "renewable + natural capital" model is being developed or explored include Mongolia, Indonesia, Vietnam, and Australia, just to name a few.

INTERVIEW
Prof. Sergio A. Sañudo-Wilhelmy
University of
Southern California,
Department of Biological Sciences
(Marine Environmental Biology),
Department of Earth Sciences

by Bill Kentrup

While coastal canyons occupy only a minute fraction of our oceans, how are they unique when considering long-term removal of CO2 flux — from our atmosphere to the ocean depths?

Coastal canyons are a conduit from coastal regions where nutrient levels are high with the potential for increasing biological carbon sequestration to the deep ocean during downwelling and other sinking events, periods when prevailing water flow is from the surface to deep water. If the organic matter produced from CO2 makes it all the way to the anoxic (lacking oxygen) ocean sediments or if it is trapped in mid- to deep water masses, this carbon would stay there for hundreds of years, based on the average time it takes for mid- to deep ocean water masses to cycle back to the surface. Studies conducted along coastal canyons on the west coast of North America have shown high biological carbon transport rates from the surface to deep waters. There are at least 650 canyons in the world's oceans, although they cannot compete with the open ocean in terms of area. Most of the open ocean is oligotrophic — meaning low in plant nutrients — with respect to nitrogen, a required element to produce biomass. Coastal areas, on the other hand, have plenty of nitrogen and other nutrients necessary for the growth of micro and macro-algae and, hence, have the potential to grow and rapidly sink biomass, as observed in the few studies carried out in coastal canyons.

Can you say something about the permanence of CO2 sequestration?

In canyons, it will depend on the physical oceanography of each location. If the carbon is trapped in a water mass that leaves the area to combine with some intermediate waters (at mid-ocean depths), carbon could be trapped in the ocean for hundreds of years. The same would be true, with potentially much longer periods of sequestration, if the carbon makes it all the way to the sediments or areas with low oxygen levels that would slow their decomposition and the release of CO2.

How could humans take advantage of coastal canyons in the context of CO2 sequestration? What could that look like? What technology would be involved?

Most canyons are located in national waters along the coast and some fertilization could be possible, i.e., introducing sufficient limiting-nutrients to grow large volumes of plant biomass. Studies need to combine physical, biological, and chemical oceanography to identify when and where plant fertilization, carbon sequestration, and subsequent sinking are feasible. There are likely to be periods of time where most plant nutrients are available in coastal areas, leaving the potential to fertilize growth through the adding of trace nutrients and/or vitamins required for growth. However, if many of the necessary nutrients are not present at a time that coincides with net movement from the surface to the depths via canyons, then the addition of trace nutrients, vitamins, or anything else limiting growth would not increase the biological uptake of CO2.

Have humans successfully tested or operationalized some of these concepts to date?

Yes and no. There have been limited studies on nutrient fertilization of plant plankton (e.g., iron) in the open oceans, as well as unintended nutri-ent fertilization (nitrogen) in coastal areas. Again, physical, biological, and chemical variables all affect the amount of plant plankton growth and the implied potential for sinking carbon. However, while research has shown the natural capacity coastal canyons have in terms of carbon transport to ocean depths, there have not been active nutrient fertilization studies within coastal canyons areas as far as I am aware. The potential for us to benefit from and utilize coastal canyons in our fight against climate change may be significant, though detailed research would need to be conducted to determine the optimal model, including operational and technological solutions that would come into play.

What are the drawbacks and potential mitigants from an environmental perspective, considering potential trophic issues, ocean acidification, and the likes?

Any addition of nutrients could potentially produce harmful algae blooms (HABs), which, depending on the species of algae, are toxic to certain marine life and/or humans. However, if the nutrient fertilization program is well designed and coincides with water sinking events in coastal canyons, most of the toxic phytoplankton would be transported and trapped in deep waters, with limited ability to impart their toxicity. Of course, as is partly determined by wind direction and strength, and other physical processes as well as topographic features, it is not entirely predictable; hence, even well designed programs would not be risk-free. Also there is water exchange between coastal waters and deep ocean canyons on the east coast of the US that is not caused by wind-driven downwelling. Oceanic circulation is complex, and there are many physical mechanisms that would need to be studied in order to design an optimal ocean carbon sink program.

What are the barriers and what would be needed in terms of technology, capital, policy, and other resources to move this concept along, from where it is now to becoming a major feature in global efforts to address climate change?

I do not know about this one. We discussed the need for research on the physics, biology, and chemistry. National policy may be an issue, depending on where projects are being implemented. Because coastal canyons tend to be in national waters, some countries might be more positive about this approach, given activities in national waters are not affected by the international dumping commission rules. Those countries could use the submarine canyons to offset their carbon balance mandated by the Paris Agreement. If we were to imagine what a coastal canyon carbon sink project could look like, we could also think about what might be needed to get from study to proof-of-concept to large-scale implementation. One model that could be considered would be coastal aquaculture (e.g., seaweed, plankton, grasses) — we could harvest the beneficial elements and "flush" the residue biomass down through coastal canyons to the deep ocean during downwelling or other sinking events.

I understand that in various countries in Asia, inland and coastal aquaculture have been going on for decades, maybe centuries. Likewise, the collaboration between the East and West in oceanography has been quite productive over the years. Perhaps as a start, what is needed is a working group of advanced industrial-scale agriculturalists and coastal oceanographers. Some discussions over the "art of the possible" between these groups could give some hints as to potential workable models and the potential technology and capital that might be required.

What would be your "call to action" to the carbon markets?

The recent report of the US Academy of Sciences recommends that we start working on blue carbon sequestration involving oceanic fertilization. I believe this call to action by the Academy sets the general tone. This, together with a specific call to action of the carbon finance market to help convene dialog and unlimitedly fund research, pilots, and commercial-scale roll outs, could reveal one of the largest and most permanent carbon sink models available.

Endnotes

1. The amount of a substance passing across a boundary or membrane, from one area to another.
2. United States Department of Agriculture, "Considering forest and grassland carbon in land management", United States Department of Agriculture, 2017, https://www.fs.usda.gov/sites/default/files/fs_media/fs_document/update-considering-forestandgrassland-carbonin-landmanagement-508-61517.pdf.
3. The average amount of time it takes for a molecule entering a system to once again exit that same system.
4. Royal Meteorological Society, "The changing carbon cycle", MetLink, Royal Meteorological Society, https://www.metlink.org/resource/the-changing-carbon-cycle/.
5. A concept that assesses whether the emission reductions of a particular project would likely have occurred in the normal course of business or by regulation or other means, even without the prospect of potential carbon finance revenue.
6. Verified Carbon Standard, a standard for approving voluntary carbon market projects and for issuing emission reduction certificates.
7. A standard for approving voluntary carbon market projects and issuing emission reduction certificates.
8. The positive impact a project might have that extends beyond carbon, such as biodiversity, job creation, technology, and skills transfer.
9. A market framework for reducing emissions from deforestation and forest degradation.
10. A non-profit investigative journalism entity.
11. Lisa Song, "An even more inconvenient truth: Why carbon credits for forest preservation may be worse than nothing", ProPublica, 22 May 2019, https://features.propublica.org/brazil-carbon-offsets/inconvenient-truth-carbon-credits-dont-work-deforestation-redd-acre-cambodia/.
12. In carbon market parlance, the Baseline Scenario is the scenario that would reasonably be expected to occur if the carbon-reducing project is not implemented, and the Project Scenario describes what is intended to occur subject to the successful implementation of the carbon-reducing project. For many carbon projects, the volume of carbon EFPs created by a project is calculated by the difference between the Baseline and Project Scenarios.
13. B. B. Cael, Kelsey Bisson, and Michael J. Follows, "How have recent temperature changes affected the efficiency of ocean biological carbon export?" *Limnology and Oceanography Letters* **2** (2017) 113–118.

14. John Weier, "John Martin, 1935–1993", The Earth Observatory, NASA, 10 July 2001, https://earthobservatory.nasa.gov/features/Martin.
15. "Systems capable of universally turning code into reliable chemistry and materials processes", Professor Lee Cronin, University of Glasgow.
16. "The world's stocks of natural assets, which include geology, soil, air, water and all living things. It is from this Natural Capital that humans derive a wide range of services, often called ecosystem services, which make human life possible". (Convention on Biological Diversity)
17. Bloomberg Law, "Climate tech startup Cultivo adds Mintz Levin lawyer as first GC", https://news.bloomberglaw.com/business-and-practice/climate-tech-startup-cultivo-adds-mintz-levin-lawyer-as-first-gc.
18. Business Wire, "Carbon opportunities fund launches first-of-its-kind investment platform to issue tokenized carbon credits", https://www.businesswire.com/news/home/20220817005164/en/Carbon-Opportunities-Fund-Launches-First-of-its-Kind-Investment-Platform-to-Issue-Tokenized-Carbon-Credits.
19. Bloomberg Law, "Climate tech startup Cultivo adds Mintz Levin lawyer as first GC".
20. CB Insights, "About Cultivo", https://www.cbinsights.com/company/cultivo-land-pbc.

Chapter 9
SEQUESTRATION

by Moon K. Kim
and Rachel Fleishman

ANALYSIS

by Moon K. Kim

Carbon is the building block of the modern world. It is the constitutive material of the fuels, chemicals, and materials that make our life possible as well as of all life on Earth. The Earth's carbon cycle is complex, with many inputs, outputs, and interconnections. Figure 9.1 summarizes the estimated stocks and net flows of carbon dioxide (CO2).

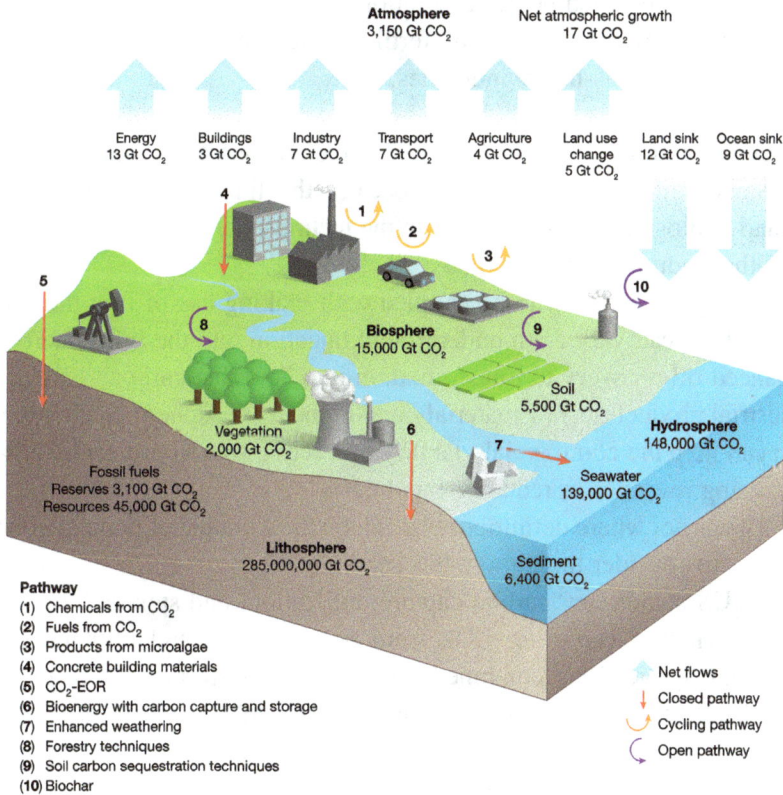

Figure 9.1. Carbon stocks and net flows with various utilization and removal pathways.[1]

Carbon sequestration can be thought of in two steps. The first is to capture CO_2,[2] and the second is to store the captured CO_2 for an extended period of time.

Carbon capture has two categories based on the source of CO_2. The first is ambient air, counting toward CO_2 removal or negative emissions. And the second is large point-sources for emissions reduction. These sources include, but are not limited to:

- natural gas processing,
- oil and petrochemicals refining,
- combustion-based power generation,
- the production of bioethanol, fertilizer, steel, cement, and
- other emissions-heavy industries.

Once captured, CO_2 can be stored in the biosphere (natural sinks such as plants, soil sediments, and the ocean), the lithosphere (deep underground geological formations), or in the technosphere (building materials and other industrial products).

Storage of carbon can be coupled with making use of it at the same time, which can vary from underground injection into the lithosphere for enhanced oil recovery to transformation into "durable carbon"[3] stored in industrial products as raw material input in the technosphere. Such utilization can help the commercial viability of a sequestration project by either increasing revenues or reducing input costs.

This is not where definitions end. There are a couple of popular abbreviations to consider throughout this section.

CCUS stands for "carbon capture, utilization, and storage". In other terms, it means capturing CO_2 from point-sources and utilizing it in industrial applications combined with long-term storage. CCS, on the other hand, stands for "carbon capture and storage", covering carbon capture from point-sources for dedicated geological storage without the economics-enhancing utilization step. These two are rather close, and we might sometimes refer to them collectively with the CC(U)S short form. There is also CDR, which simply stands for "carbon dioxide removal", used interchangeably with "negative emissions" and applied to capturing CO_2 from ambient air for permanent storage.

Sequestration of CO2 is essential for hard-to-decarbonize industries (such as oil and gas, chemicals, steel, and cement) to achieve their climate goals and for limiting global warming to 1.5–2°C by directly removing excess CO2 being accumulated in the Earth's atmosphere by human activities. Importantly, in order for carbon sequestration to be an effective decarbonization tool, all types of CO2 storage need to be permanent or long enough to buy time until the next generation of carbon removal technologies becomes widely adopted. This implies timelines requiring decades to centuries.

Carbon can be stored near-permanently in the lithosphere, mineralized in the built and natural environments, or deep ocean. Assuming no leakage, carbon storage elsewhere in the biosphere lasts in the order of decades to centuries, while industrial utilization as CO2-derived fuels and chemicals store carbon only for days to months. As the previous chapter was dedicated to forestry and oceans, in this one, we will be focusing primarily on industrial sequestration and methods with significant human involvement.

Also, as the Asia-Pacific region lacks significant developments in industrial sequestration projects, we will provide a technical overview, a global snapshot, and some relevant examples from the region.

PAST

Nature absorbs more than half of anthropogenic CO2 emissions via land and ocean sinks. Preserving and restoring the world's forests and wetlands — in other words relying on photosynthesis to capture atmospheric CO2 and then storing it in the biosphere — have been considered among the easiest, lowest-cost ways of CO2 removal or CDR. The total cost of carbon sequestration via forestation[4] is estimated to be lower than USD 10 per tonne[5] of CO2 removed, without accounting for forests' direct economic benefits, including the generation of food, fiber, and carbon credits.

Yet, this biological carbon removal can be easily reversed when the ecosystem is disturbed. To prevent the release of stored carbon back into the atmosphere, forests must be managed sustainably, and the wood must be preserved from incineration or decomposition. One possible solution is by utilizing it in construction or biochar.[6]

As discussed in the previous chapter, forestation and sustainable forest management have historically been the most popular methods for large corporations to offset greenhouse gas (GHG) emissions from their operations. However, the efficacy of such forest offsets has been recently called into question when prominent environmental groups were found to have sold large amounts of corporate carbon credits for preserving already well-protected trees in safe ecosystems.[7] Budgeting carbon sequestration via biological pathways can also be challenging. Effects of climate change on forest productivity are uncertain, with higher concentration of atmospheric CO_2 stimulating photosynthesis and thus increasing growth rate of some tree species in certain regions while changes in water cycle (especially droughts) negatively impact others in their growth rates and resistance to pests and diseases.

The first large-scale CCS project was at Norway's Sleipner Vest offshore gas field in the North Sea, where StatoilHydro[8] was already separating CO_2 from produced gas during natural gas processing. Since its commissioning in 1996, the Sleipner project has captured 1 Mtpa (megatonne[9] per annum) of CO_2 and stored it in a subsea saline aquifer.[10] A carbon tax on offshore oil and gas activities introduced by the Norwegian government in 1991 made the project commercially viable.

Virtually all of the early commercial CC(U)S projects until the 2000s came from natural gas processing plants. During the production process itself, CO_2 must be stripped out to meet market requirements for natural gas, as well as to avoid freezing of CO_2 that can damage processing equipment.

PRESENT

Despite some drawbacks, nature-based CDR solutions must still be a major part of decarbonization tools that we deploy now and in the future.

Restorative biosequestration not only serves as the lowest-cost, productive carbon removal pathway but also provides added benefits of improving biodiversity, water and air quality, and flood control, all of which are important factors in climate resilience. Forests and oceans are also sources of livelihood for local communities worldwide.

There are numerous innovative startups around the world looking to address challenges facing nature-based climate solutions. Businesses in development and technologies in commercialization range from the more scientific to the more mundane in Asia-Pacific and around the world. For the former, a good example is Singapore's Double Helix Tracking Technologies, which uses DNA-based testing of forest products to independently verify claims in species and origin. Such independent verification ensures a sustainable supply chain. For the latter, there are Dendra Systems in the United Kingdom (UK). The company's drones fire seed pods into the ground to automate forestation faster and at a lower cost than current methods.

However, let us focus on the other carbon sequestration pathways with commercially viable technologies. The two main areas are CC(U)S and soil carbon sequestration; the latter addresses regenerative agriculture.

Carbon Capture (Utilization) and Storage

In CC(U)S facilities operating today, CO_2 is captured via industrial separation[11] or post-combustion separation from the flue gas[12] produced by fossil fuel combustion. Captured CO_2 is then compressed and transported to storage or utilization sites by pipeline, ship, rail, or tanker truck, in decreasing order of usage. Pipeline transportation, where available, is usually the cheapest option.

Economies of scale with today's carbon capture technologies favor larger-scale applications like natural gas processing plants,[13] which have large volumes of CO_2 in high concentration. Recently commissioned projects and those in development are more varied (see Figure 9.2), with lower concentration sources such as power generation and production of steel or cement growing and emerging, respectively.

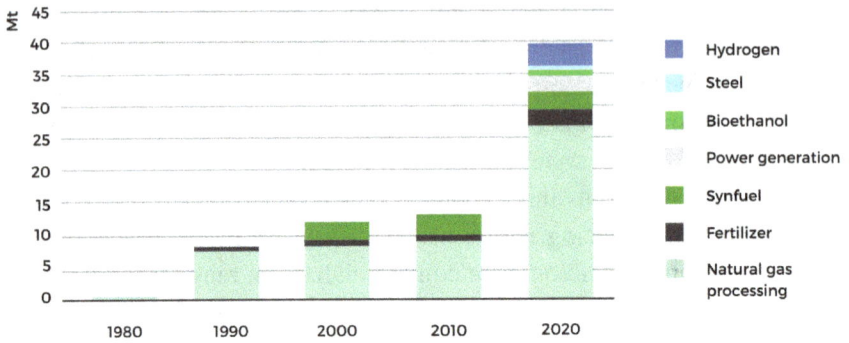

Figure 9.2.   Global CO2 capture capacity at large-scale facilities by source.[14]

Suitable geological storage sites are selected based on certain geophysical conditions[16] and the absence of nearby fault planes[17] to prevent CO2 from leaking into the atmosphere. There is more than enough carbon storage capacity worldwide[18] to meet the 220 gigatonnes[19] (Gt) of carbon storage needed over the next 50 years, as forecast by the IEA Sustainable Development Scenario (see Figure 9.3).[20]

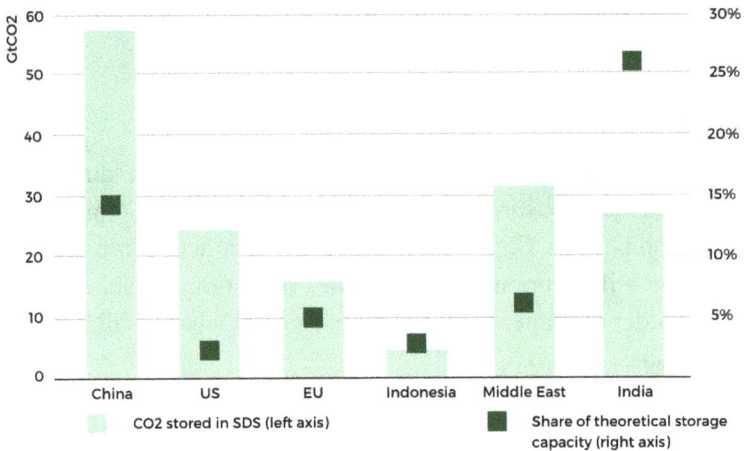

Figure 9.3.   Cumulative CO2 storage in the Sustainable Development Scenario (SDS) and share of theoretical storage capacity, by region.[15]

Unlike CCUS, CCS has no associated revenue streams from industrial carbon utilization to offset costs. The carbon price needed to make CCS economically viable is estimated at USD 40–80/tonne in 2020 and USD 50–100/tonne by 2030.[21]

CCUS — Enhanced Oil Recovery

Among the many industrial applications[22] of CO2, utilization in Enhanced Oil Recovery (EOR) is the most common today. Pressurized CO2 is injected into older oil fields to improve oil's mobility and increase its flow toward the production wells, boosting oil production while displacing oil that was once trapped in the geological formations. Most commercial EOR projects recycle the CO2 that returns to the surface, resulting in >99% of the injected CO2 being permanently stored over the life of the project.[23]

The price paid for CO2 in EOR[24] is typically indexed to the oil price, which can determine economic viability. For example, the drop in oil prices during the COVID-19 pandemic severely impaired the Texas-based Petra Nova Carbon Capture[25] facility with EOR[26] and led to the suspension of its operations in March 2020. The activity will be restarted when economic conditions improve.

The vast majority of current EOR operations use terrestrial CO2 dug up from natural underground reservoirs (versus carbon captured in post-industrial processes), exacerbating the carbon footprint of oil production. Producers blame this practice on the limited availability of low cost, reliable sources of anthropogenic CO2 close to oilfields. EOR can also be problematic in terms of political economy, as it enables the recovery of more oil, potentially extending dependence on fossil fuels. As an on-ramp for reducing costs and scaling up of CCS, more widespread adoption of EOR can still help make progress on climate change, as long as the CO2 used is anthropogenically sourced.

CC(U)S Implementation

In 2020, there were 28 commercial CC(U)S facilities in operation around the world with a total CO2 capture capacity of 40 Mtpa.[27] The United Nations (UN) Framework Convention on Climate Change estimates that the planet needs somewhere between 200 and 400 Mtpa in 2020 to

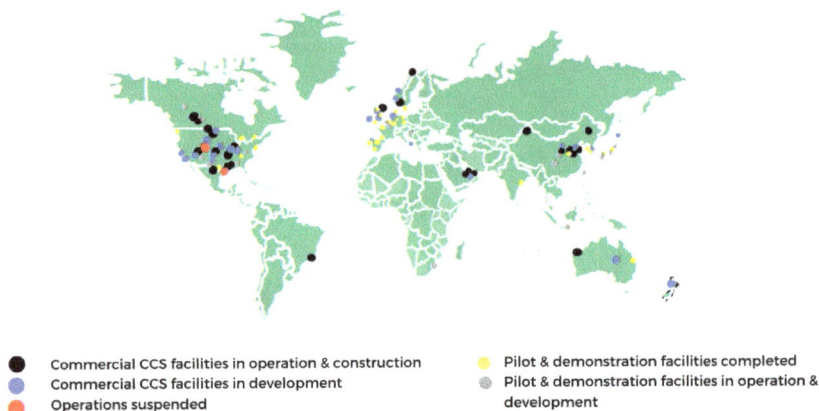

● Commercial CCS facilities in operation & construction   ● Pilot & demonstration facilities completed
● Commercial CCS facilities in development   ● Pilot & demonstration facilities in operation &
● Operations suspended   development

Figure 9.4.   World map of CC(U)S facilities at various stages of development.[28]

1.5 Gtpa (gigatonnes per annum) by 2030 and 6.3 Gtpa by 2050 to reach the temperature goal. There are 37 new commercial CC(U)S facilities in construction and advanced or early development (see Figure 9.4). If all these projects proceed as planned, adding around 75 Mtpa,[29] the total global capacity would nearly triple to 115 Mtpa by the mid-2020s (still significantly below the UN's projected need).

Facilities in operation are concentrated in the United States (US) (half of the total sum), with Australia, Brazil, Canada, China, Qatar, Saudi Arabia, and the United Arab Emirates (UAE) commissioning projects in the last decade (see Table 9.1). The US also dominates in upcoming projects,[30] mostly due to the favorable tax credit policy and access to CarbonSAFE[31] — CO_2 storage hubs supported by the US Department of Energy. Momentum is growing in Europe, where Ireland, the Netherlands, Norway, and the UK are also developing projects supported by the European Green Deal and Climate Law. Commercial facilities in construction or in advanced development in Asia-Pacific span across Australia, China, New Zealand, and the UAE (see Table 9.2). Malaysia[32] and South Korea are in the early planning-development stages.

The overwhelming majority of the commercial CC(U)S operations installed to date utilize EOR. Only 6 out of the 28 operating facilities are CCS projects,[33] depositing CO_2 in dedicated geological storage without the EOR utilization step. Today, the adoption of CCS is picking up speed in

Table 9.1. List of commercial operating CC(U)S facilities in Asia-Pacific.[34]

Facility Title	Status	Country	Operation Date	Industry	Capture Capacity (Mtpa max)	Capture Type	Storage Type
Sinopec Zhongyuan Carbon Capture Utilzation and Storage	Operational	China	2006	Chemical production	0.12	Industrial Separation	Enhanced Oil Recovery
Uthamaniyah CO2-EOR Demonstration	Operational	Saudi Arabia	2015	Natural gas processing	0.80	Industrial Separation	Enhanced Oil Recovery
Karamay Dunhua Oil Technology CCUS EOR	Operational	China	2015	Chemical production methanol	0.30	Industrial Separation	Enhanced Oil Recovery
Abu Dhabi CCS (Phase 1 Emirates Steel Industries)	Operational	UAE	2016	Iron and steel production	0.80	Industrial Separation	Enhanced Oil Recovery
CNPC Jilin Oil Field CO2 EOR	Operational	China	2018	Natural gas processing	0.60	Industrial Separation	Enhanced Oil Recovery
Gorgon Carbon Dioxide Injection	Operational	Australia	2019	Natural gas processing	4.00	Industrial Separation	Dedicated Geological Storage
Qatar LNG CCS	Operational	Qatar	2019	Natural gas processing	2.10	Industrial Separation	Dedicated Geological Storage

Table 9.2.   Commercial CC(U)S facilities in construction or development across Asia-Pacific.[35]

Facility Title	Status	Country	Operation Date	Industry	Capture Capacity (Mtpa max)	Capture Type	Storage Type
Yangchang Integrated Carbon Capture and Storage Demonstration	In Construction	China	Delayed to 2020s	Chemical production	0.41	Industrial Separation	Enhanced Oil Recovery
Sinopec Shengli Power Plant CCS	Early Development	China	2020s	Power generation	1.00	Post-combustion capture	Enhanced Oil Recovery
Korea-CCS 162	Early Development	South Korea	2020s	Power generation coal-fired	1.00	Under evaluation	Dedicated Geological Storage
Sinopec Qilu petrochemical CCS	In Construction	China	2020–2021	Chemical production	0.40	Industrial Separation	Enhanced Oil Recovery
Sontos Cooper Basin CCS Project	Advanced Development	Australia	2023	Natural gas processing	1.70	Industrial Separation	Dedicated Geological Storage
Project Pouakai Hydrogen Production with CCS	Early Development	New Zealand	2024	Hydrogen Production and Power Generation	1.00	Industrial Separation	In evaluation
Abu Dhabi CCS Phase 2: Natural gas processing plant	Advanced Development	UAE	2025	Natural gas processing	2.30	Industrial Separation	Enhanced Oil Recovery

the West, thanks to the recent refocus of both the public and private sectors on achieving net zero GHG emissions. Out of the 32 projects in development in the West, 23 are expected to be CCS.[36] In the Asia-Pacific region, one commercial CCS project is currently in advanced development in Australia, and another is in early development in South Korea.

Other CC(U)S Developments in Asia-Pacific

China has recently built 12 large-scale CC(U)S projects of 15.2 Mtpa combined capacity in the pipeline,[37] in addition to the three facilities in construction/development listed in Table 9.2. CC(U)S is an essential part of China's path to carbon neutrality by 2060 and is estimated to provide 1.5–2.7 Gtpa of emissions reductions by 2050, as shown in Figure 9.5. This rapid scale-up will be facilitated in part by creation of CC(U)S hubs in major industrial regions to provide shared pipeline and storage infrastructure for carbon capture from various industrial sources.

The Gulf Cooperation Council (GCC) region produces around 25% of the annual global oil output and has an accessible underground carbon storage potential of 5–30 Gt.[38] The region also needs significantly more CO2 to replace natural gas that has been heavily utilized for EOR-assisted oil production. Considering these, along with abundant natural gas resources and excess production capacity, as well as stated ambitions for developing a clean hydrogen export industry, countries of the GCC may

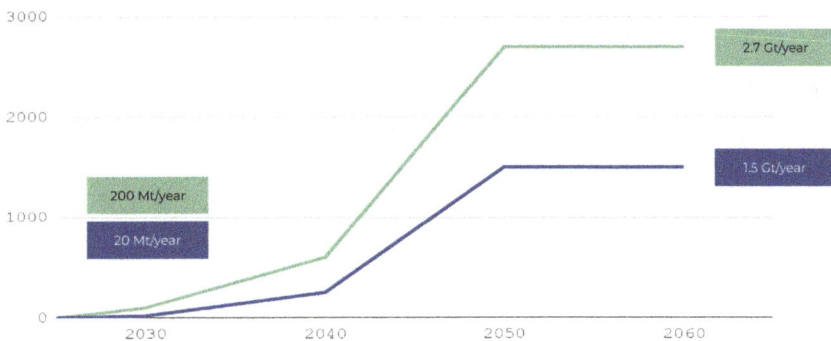

Figure 9.5. Potential CC(U)S deployment trajectories in China (Mtpa).[39]

yet emerge as significant actors in CCUS despite limited development to date.

The Abu Dhabi National Oil Company, the GCC region's most active player in CCUS, has pledged to cut 25% off its GHG intensity by 2030 and is accelerating CCUS investment to reach 5 Mtpa capacity in the next ten years. In addition to Phases 1 and 2 plants at the Al Reyadah site listed in Tables 9.1 and 9.2, the company is planning a Phase 3 facility to capture another 1.9 Mtpa from the Habshan-Bab natural gas processing plant for EOR.

Lastly, interesting multinational CC(U)S projects are afoot in Asia-Pacific. For example, Australia's Transborders Energy launched a collaboration with oil and gas, utilities, and transport companies as well as research organizations across Australia, Japan, Norway, and the US[40] in late 2020 to consider co-developing an offshore CCS Hub project called the "deepC Store Project". This project aims to capture CO2 from industrial sources (liquefied natural gas plants in particular) in Australia and Japan and ship liquid CO2 from capture sites to a floating storage and injection hub facility in offshore Australia to store in subsurface wells. In mid-2021, Japan's Ministry of Economy, Trade and Industry formed the Asia CCUS Network with over 100 companies[41] in the ten Association of Southeast Asian Nations (ASEAN) members, Australia, Japan, and the US, to share technological know-how and to jointly pursue CC(U)S projects along with CO2 liquefaction for shipping across the Asia-Pacific region.

Bioenergy with Carbon Capture and Storage

Unlike CCS, which reduces emissions, Bioenergy with Carbon Capture and Storage (BECCS) aims to achieve negative emissions. The growth of plant biomass absorbs atmospheric CO2 via photosynthesis, and biogenic CO2, released from the utilization of such biomass or biomass-derived products, is counted as a net zero emission in most GHG accounting methodologies. Capturing and permanently storing biogenic CO2, similar to the broader CCS already discussed, results in a net reduction of CO2 from the atmosphere.

As seen in Figure 9.6, BECCS encompasses a variety of

- biomass feedstocks, including wood, dedicated energy crops, agricultural and forest residues, and municipal solid waste,
- methods of energy conversion, such as combustion, fermentation, and anaerobic digestion, and
- end-uses of bioenergy, like the production of heat or steam for power generation, transportation fuels, and industrial applications.

BECCS projects developed to date capture CO2 from bioethanol production or biomass-fed power generation. There are 11 operating BECCS facilities worldwide, collectively capturing less than 2.6 Mtpa of CO2, and all but one are demonstration-scale projects ranging from 3 ktpa (kilotonnes[42] per annum) to 290 ktpa in capture-capacity.[43] Archer Daniels Midland-owned Illinois Industrial CCS facility, commissioned in 2017 with 1 Mtpa capture capacity and located at the Decatur corn-to-ethanol processing plant, is the largest and the only project with dedicated geological carbon storage. The other ten projects rely on EOR or use the captured CO2 for industrial applications without storing.

The vast majority of the facilities are based in North America and Europe. They treat the fermentation byproduced gas streams from ethanol

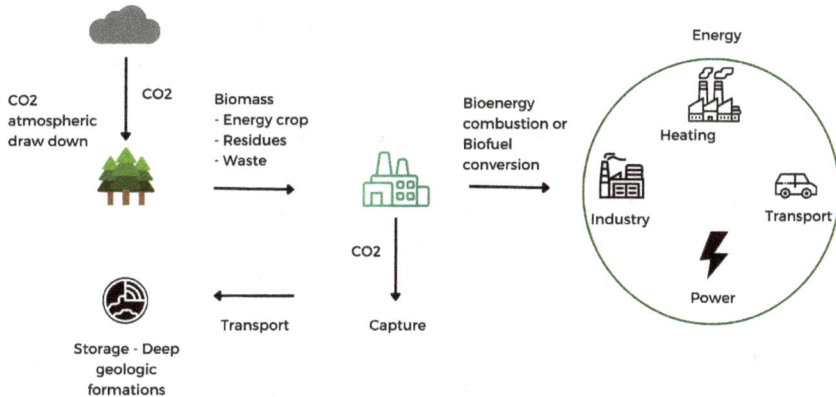

Figure 9.6. Schematic of bioenergy with carbon capture and storage.[44]

plants, consisting purely of CO2 and water. These are dehydrated and compressed for transport and storage. This makes bioethanol plants the most economically attractive option for BECCS. However, subsequent combustion of the bioethanol releases CO2, discounting the CDR benefit of BECCS.

The two facilities not sourcing from ethanol plants are based in Japan. The first is a 3 ktpa Saga City waste-to-energy project commissioned in 2016. The second is a 180 ktpa Mikawa post-combustion capture project, commissioned in 2020, located at a 49 MW biomass power plant.

The cost of BECCS varies widely, ranging between USD 15 and 400 per tonne of CO2 removed, depending on the sector. BECCS involving bioethanol production is found to be the least expensive at USD 20–175 per tonne avoided.[45]

Among industrial CDR technologies, BECCS is the most commercially viable. The individual technologies utilizing biomass as the source of energy (e.g., biofuels, biomass power generation, waste-to-energy[46]) are mature, with worldwide commercial application, while CCS is an established technology with broader deployments in development or planning. Because of this relative maturity, climate change integrated assessment models firmly rely on BECCS as the most widely used negative emissions technology to draw CO2 from the atmosphere at the gigatonne scale from 2030 to 2100 to meet climate targets. Climate models that assume more aggressive reductions in current emissions require fewer CDR technologies and thus less BECCS. Even in the most aggressive scenarios, 2–3.3 Gtpa of BECCS is needed by the end of the century.[47]

There is significant uncertainty about the scale of this contribution even at the lower end of the scenarios, as the supply of sustainable biomass may be constrained by the availability of land, water, and fertilizer. Meeting the upper bounds of the BECCS targets would require three times the total cereal production, twice the agricultural water demand, and 20 times the use of nutrients in the world on an annual basis.[48]

The potential for future large-scale implementation of BECCS should not be relied on as an alternative to achieving rapid emissions reductions across all sectors today. It should be considered an essential complement to the broader deployment of CC(U)S technologies by heavy industries to reach net zero emissions. Further, the life cycle of each BECCS project needs to be assessed and monitored to ensure that genuine negative emissions are achieved based on sustainable land use and management as well as the sourcing of biomass feedstock.

Scaling Challenges in CC(U)S

CC(U)S installations have been steadily scaling up but have not been broadly adopted by emissions-heavy industries. Many planned projects have been delayed or stalled at the gate due to commercial considerations, including, but not limited to the heavy CAPEX-intensity, access to financing, and the integration and execution of complex multi-party negotiations. The lack of consistent policy support such as incentives or emission penalties has also been a hindrance. Meanwhile, there are continuing engineering and technological advancements to improve the economic attractiveness of CC(U)S toward larger-scale adoption, especially concerning cost and energy-efficient CO2 capture.

For CC(U)S to grow beyond the realm of state-owned enterprises and large multinational corporations, modular containerized carbon capture plants leveraging the economies of manufacturing scale will be needed for economic application at a smaller scale. Such size reduction will also support BECCS since a significant portion of the world's biogenic CO2 emissions sources is of modest-sized waste-to-energy facilities, or pulp and paper plants.

Norway's Aker Carbon Capture has developed a modular product, "Just Catch",[49] using the same chemical absorption process and amine-based solvent proven at larger capture plants. The company is scheduled to deliver the first Just Catch system to the Twence waste-to-energy plant in the Netherlands by late 2021 for 100 ktpa of post-combustion carbon capture. Another Norwegian company Compact Carbon Capture (3C) developed a compact modular capture technology using rotating beds of solvents instead of large static columns in a traditional capture plant. Baker Hughes acquired the firm in late 2020. 3C's technology enables up to 75% reduction in size and lower capital requirements. Baker Hughes will further develop and scale the technology, leveraging its expertise in rotating equipment to offer low-cost CO2 capture solutions for global deployment, including the Asia-Pacific region.

The carbon capture process makes up the largest portion of the overall cost and energy requirement of a CC(U)S system because separating CO2 from a mixture of other gases and pollutants is costly and energy-intensive. Although separation is fairly straightforward for industrial waste streams with high concentrations of CO2, such as in natural gas processing or ammonia production, it is costlier for those with low concentrations (see Figure 9.7). A coal-fired power plant retrofitted for CO2 capture can require about 25% more fuel to generate the same amount of electricity as one without such retrofit.

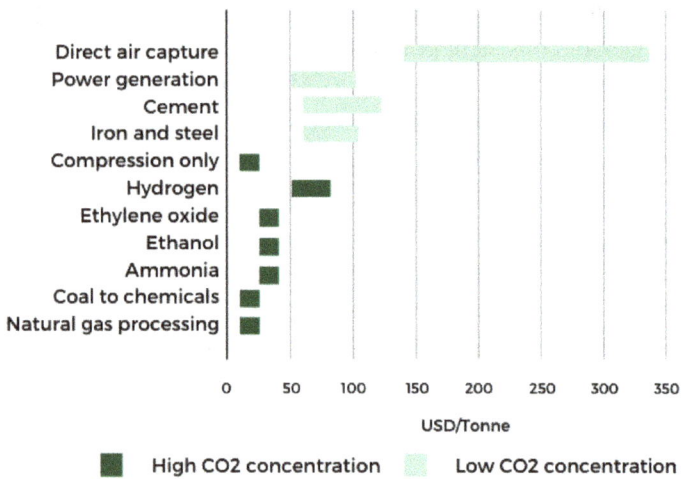

Figure 9.7. Levelized cost of carbon capture by sector and initial CO2 concentration, 2019.[50]

Chemical absorption of CO2 using amine-based liquid solvents is currently the most widely used carbon capture process but demands a lot of thermal energy to release CO2 back from the solution and regenerate the solvent for reuse. Solvent degradation[51] produces chemical waste that requires safe disposal and also adds to the costs.

Point-source Carbon Capture Technology Startups

Advanced solvents are being developed to address these challenges. For example, UK-based Carbon Clean Solutions has developed a proprietary liquid solvent showing significantly improved absorption efficiency with 20–40% lower energy requirement and less solvent degradation while resisting plant corrosion. The technology is proven at a demonstration scale with more than ten installations across Europe, the US, and India. These include a 63.5 ktpa facility that captures CO_2 from coal-fired boiler flue gas at the Tuticorin baking soda/soda ash plant in Tamil Nadu, which used to purchase CO_2 in the past. Carbon Clean received venture capital funding in 2020 to execute an existing pipeline of global projects and invest in developing modular containerized solutions to achieve USD 30 per tonne cost of carbon capture.[52]

Physical adsorption of CO_2[53] has been used for many years in industrial gas separation systems. Still, conventional temperature swing adsorption (TSA) processes for high-volume carbon capture present challenges with long cycle times and high equipment costs. Canada's Svante has developed a proprietary adsorbent filter,[54] which uses significantly less energy to regenerate, combined with a rotating rapid TSA process for carbon capture from dilute industrial flue gas. The startup's capture system is small enough to retrofit by directly connecting to a flue stack and costs half the capital expense of a traditional capture plant. Svante aims to reach USD 15 per tonne capture cost and has been operating a 10 ktpa pilot plant at Husky Energy's heavy oil production site in Saskatchewan since 2019 and a demonstration facility at Lafarge's cement production plant in British Columbia since early 2021. Canadian integrated oil and gas company Suncor Energy invested in Svante in March 2021 to develop low-carbon fuels and blue hydrogen.

LanzaTech, originally from New Zealand, takes an entirely different approach to point-source carbon capture utilizing microbial fermentation. Carbonaceous industrial waste gas is fed to proprietary microbes in a bioreactor recycling the carbon into fuels and commodity chemicals.[55] In 2020, the startup established LanzaJet JV with Suncor and Mitsui to produce sustainable aviation fuel, starting with a 10-million gallon-per-year demonstration plant in Georgia that is expected to be completed by 2022. LanzaTech has received more than USD 300 million in venture capital and corporate funding, including USD 30 million investment in late 2021 from ArcelorMittal, which plans to deploy the startup's carbon capture and reuse technology at its steel plant.

Soil Carbon Sequestration

Soil is a major carbon sink, storing it as inorganic elemental carbon and carbonate minerals or as organic carbon contained in soil organic matter, which is concentrated in the topsoil.[56] Soil organic carbon exists in dynamic equilibrium where it is lost by respiration and decomposition of organic matter as well as erosion and leaching, and its quantity depends on the soil type,[57] climate (by regulating plant productivity), and management practices (land and soil management or crop strategy).

Global soil organic carbon stocks have been declining due to the conversion of natural ecosystems to farmland, overgrazing, and other prevailing agricultural practices. Land-use change is estimated to have caused up to 60% loss of soil carbon to the atmosphere in temperate regions and 75% or more in the tropics.[58] This loss, however, can be reversed (see Figure 9.8) via afforestation, reforestation, and agroforestry, all of which are discussed in the previous chapter, "Forestry & Oceans".

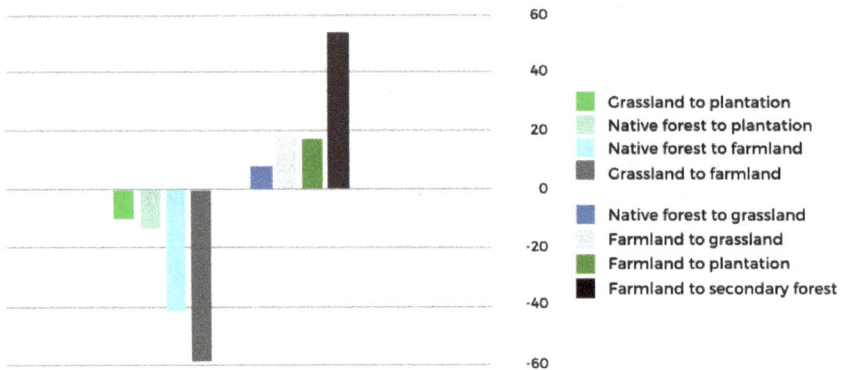

Figure 9.8.   Changes in carbon stocks (%) after land use change.[59]

Another pathway for soil carbon sequestration is regenerative agriculture, which generally aims to increase the input of organic carbon materials into the soil and minimize disturbance and erosion that increase soil organic carbon loss. A variety of sustainable agricultural techniques are used in combination for regenerative farming, and best management practices include:

  i) reduction or elimination of mechanical tillage,
 ii) erosion control via contour plowing or terracing,

iii) use of crop residues as surface mulch,
iv) incorporation of cover crops and diverse crop rotations,
v) application of organic waste as fertilizers (e.g., compost, manure, biosolids) along with reduced use of chemical inputs and pesticides,
vi) planned grazing (versus continuous grazing) for better grassland management, and/or
vii) complex farming systems such as mixed crop-livestock that efficiently use resources, enhance biodiversity, and mimic the natural ecosystems.

There are numerous benefits of regenerative agriculture in terms of long-term farm economics, ecological services, and resilience to climate change. A higher level of organic matter improves soil quality through increased retention of water and nutrients, resulting in greater farm yield as well as less synthetic fertilizer usage (and the accompanying emissions of nitrous oxide),[60] and boosts soil health and resilience against extreme weather. It also enhances soil structure and reduces erosion and nutrient leaching, leading to improved water quality in both inland aquatic and coastal ecosystems.

Scaling Challenges in Soil Carbon Sequestration

What is unclear is its practical potential for large-scale carbon sequestration. Maintaining enhanced soil organic carbon levels is difficult, especially for the long term. If regenerative practices are discontinued (e.g., a farmer stops the no-till method, changes crop strategy, sells his land, or suffers an extreme weather event that erodes his soil), sequestration is easily reversed. Consistent definitions or standards for carbon sequestration via regenerative farming have not been established, but at least eight companies ranging from startups to large multinationals have launched agricultural carbon credit programs since 2019. While a few companies, such as Boston-based Indigo Ag, hold themselves to the emissions offset protocol's additionality and permanence principles, others are operating under dubious criteria with little public oversight. For example, some sell carbon credits generated from farms that have already been using low-carbon methods for many years. Several programs do not require long-term agreements to ensure that the farmers continue regenerative practices for the decades necessary to remove CO_2 from the atmosphere permanently.

High-quality standards and oversight organizations are starting to emerge, such as The Climate Action Reserve's Offset Project Registry for California's Cap-and-Trade Program, to minimize the risk of reversal by requiring farmers or credit developers to monitor and report on projects for a hundred years in some cases and pay for an insurance policy against reversals.[61]

On the other hand, following proper protocols for earning carbon credits from regenerative farming can be costly. In addition to buying new equipment and/or procuring new inputs when transitioning to regenerative methods, periodic measurement, reporting, and verification (MRV) to substantiate climate benefits is expensive and time-consuming, typically involving third-party companies to measure soil carbon levels, observe farm management practices, and provide documentation. This conventional MRV process has been shown to make economic sense only in a high carbon price environment (>USD 100/tonne) with a large project size (>1,000 hectares[62]), costing more than the potential carbon credit revenue for an average grassland in California.[63]

Independent soil carbon verification platform Regrow Ag, launched in early 2021 by the merger of an Australian agronomic analytics company with a satellite data-enabled soil health modeling startup based in the US, aims to empower the agricultural industry to adopt, scale, and monetize sustainable practices. Using scientifically vetted soil and carbon modeling and a publicly accessible aggregate map of regenerative farming operations based on current and historical satellite imaging since 2005, the startup offers MRV for carbon sequestration with accurate, low-cost soil insights without frequent sampling for on-the-ground reference data. Regrow is analyzing over 150 million acres in North America for customers, including Cargill, Bayer, and General Mills, and has expansion plans for Latin America, Australia, and Europe.

Risk perceptions of farmers also remain a challenge. Many growers manage their operations in a way they feel is optimized and may be reluctant to make even small changes lest they lead to yield loss. Time, consensus on high-quality applications, cost-effective MRV methodologies, and policy and finance support will be needed to shift the critical mass of farmers' cultivation philosophy to take up regenerative practices on a meaningful scale. Fintech solutions, such as product performance assurance and a warranty-backed crop plan provided by Iowa-based startup Growers Edge, may help speed up this otherwise slow adoption curve by reducing risks in embracing agtech innovations and sustainable transitions.

Regenerative agriculture has largely been a US phenomenon as the term was popularized by an organic farmer there. Still, significant parts of its practice are found in land management and farming methods around the world that have been observed for generations. Regenerative pastoralism[64] in Australia and natural farming[65] of Japan are good examples for Asia-Pacific. As interests rise, a myriad of educational programs, local farmer networks, and service groups have emerged to encourage adoption and standardization toward enduring soil carbon sequestration that can generate carbon credits. Anecdotes of adoption in other parts of Asia-Pacific include The Nature Conservancy-supported projects in arid potato-growing regions of China and Thai coconut farms driven by a consumer brand selling coconut-based organic drinks.

Asia has about 350 million smallholder farmers, constituting 78% of the global total,[66] and this makes education and coordination for regenerative farming an additional challenge where microfinanciers and impact investors may be able to play a significant role.

FUTURE

Direct Air Capture

Most 1.5–2°C pathways in climate change integrated assessment models heavily relied on the gigatonne-scale deployment of negative emissions technologies before the mid-century, including direct air capture (DAC). DAC draws CO_2 directly from ambient air using industrial technologies, either for permanent storage and carbon removal (DACCS[67]) or for industrial utilization, such as in food processing or synthetic fuel production, which re-emits the captured carbon.

There are some key advantages to DAC that differentiate it from biological CDR mechanisms such as forestation, BECCS, and regenerative agriculture. First, it can be built anywhere in the world — near geological storage sites or CO_2 off-takers to minimize transportation costs — and/or close to sources of waste heat or renewable energy to reduce its own energy and carbon footprint. Second, DAC requires an orders-of-magnitude smaller amount of land per tonne of carbon removed and can be a net producer of water in cool and humid conditions.[68]

The main drawback of DAC is the very dilute CO_2 concentration in ambient air,[69] making the technology highly energy-intensive and costly.[70]

Today's direct air capture technologies extract CO_2 from the air using (a) hydroxide-based liquid solvent for absorption or (b) solid sorbent materials for temperature swing adsorption, and then apply heat to separate it from the media in both types of technologies. Liquid absorption systems require 900°C to release captured CO_2, whereas solid adsorption systems require only 80–120°C, for which lower-grade waste heat can be used. Table 9.3 summarizes the differences among the three leading companies developing these DAC technologies.

Almost all of the DAC plants currently operated by these three are small-scale pilot and demonstration projects, with a capture capacity ranging from 1 tpa to 4,000 tpa (tonnes per annum), and their CO_2 is sold for various industrial utilization (e.g., in the production of chemicals and fuels, beverage carbonation, greenhouse fertilization) rather than geological storage.[71]

DAC Implementation

One pilot plant in Iceland by Climeworks[72] first dissolves the captured CO_2 in a geothermal power station's reinjection water and then pumps it deep underground, where it reacts with surrounding basalt formations to form mineralized carbon (see Figure 9.9). The startup's new 4 ktpa plant in Iceland started operating in September 2021. Climeworks' current costs are still high at USD 600–800 per tonne of carbon removed, but it is working towards USD 200–300 per tonne by 2030 and half of that by the late 2030s.

Table 9.3. Status of the leading DAC companies.[73]

Company	Climeworks	Global Thermostat	Carbon Engineering
Location	Switzerland	United States	Canada
System type	Solid sorbent	Solid sorbent	Liquid solvent
Thermal energy needs	80–120°C/176–248°F	80–100°C/176–212°F	900°C/1,652°F
Thermal energy source	Non-fossil energy resources (geothermal, waste heat, etc.)	Energy resource agnostic	Natural gas with CCS
Projects	Commercial operation with 16 plants globally with a collective capacity of 2,000 tonnes of CO2 captured from air per year	Prototype plants in California and Alabama	Pilot plant in British Columbia; in the process of building facility in the Permian Basin that will be capable of 1 million tonnes of CO2 capture per year
Investments	Around USD 770 million in equity investment since founding in 2009, including CHF 600 million in the most recent round of funding Investors include GIC, Partners Group, Swiss RE, and others	Partnered with companies, including ExxonMobil, NRG, and BASF Investments from ExxonMobil, NRG Energy, financial magnates, and others	Received investments of USD 68 million in most recent round of funding in 2019 Investors include BHP, Chevron, Bill Gates, Oxy Low Carbon Ventures, and others

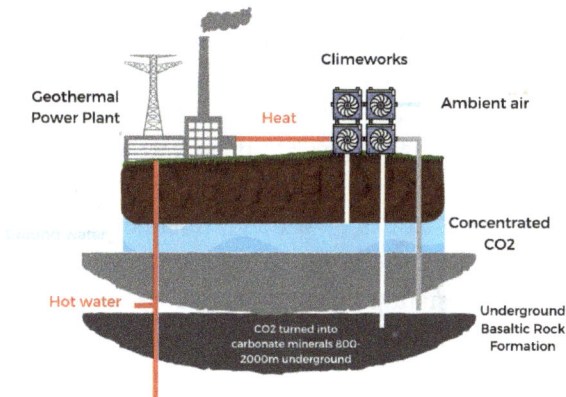

Figure 9.9. Schematic of Climeworks' DACCS project in Iceland.[74]

Carbon Engineering has partnered with Occidental Petroleum to build a 1 Mtpa DAC facility in Texas (to begin operation in 2024) for EOR in the Permian Basin. The startup is building another 1 Mtpa[75] facility in Scotland with an infrastructure developer Storegga Geotechnologies to capture emissions from hydrogen production using natural gas extracted from the North Sea and then store the captured CO2 in nearby offshore geological sites. Carbon Engineering forecasts its costs to become as low as USD 150 per tonne, including geological storage.[76]

Israel-based newcomer RepAir Carbon is building modular DAC machines based on fuel-cell electrochemistry. It claims to require much less heat to remove CO2 from its filters, reducing energy consumption for each tonne of CO2 captured by up to 67% compared to existing DAC technologies. The startup raised seed funding in Q4 2021 to advance its prototype.

Although DACCS is at a relatively early stage of development and scale-up, its advantages over nature-based carbon removal solutions present a significant opportunity. Widespread deployment of DACCS to make a meaningful climate impact by mid-century will depend on rapid cost reduction and climate policies supporting higher demand for CDR and sufficient carbon prices. According to a recent study based on the learning curves of successful technologies,[77] the DAC industry will need to grow its installed capacity by more than 300-fold to reach costs of USD 100 per tonne, and this may require government subsidies of up to USD 2 billion[78] — a surprisingly small sum in the context of global economic wealth at risk from the climate crisis.

There is a novel type of CDR technology emerging from theoretical domains, called enhanced weathering, based on a naturally occurring geological weathering process where CO2 in air and water chemically binds to minerals and turns into rocks. San Francisco-based Heirloom says their technology can induce oxides of widely available, low-cost minerals to permanently absorb CO2 from the ambient air in days (versus years in natural process). The startup's ambitious goal is to remove one billion tonnes of CO2 from the atmosphere by 2035.

Carbon Transformation

Once captured, CO2 can be transformed into "durable carbon" stored in long-lived products. Among several applications in various stages of technological and commercial maturity, incorporating captured CO_2 into concrete via mineralization has the biggest climate potential[79] and the highest commercial readiness.

The two relatively established processes are carbonation and concrete curing. In carbonation, CO_2 gas reacts with minerals in industrial waste streams to form solid carbonates, which can replace natural limestone used in concrete production. California-based Blue Planet applies this process, for instance. In concrete curing, CO_2 reacts with the cement during the curing process, resulting in similar mineralization and stronger concrete. This method is commercialized by two North American companies, Solidia Technologies and CarbonCure. CarbonCure's activities in Asia are explored in the startup case study of this chapter.

CONCLUSION

Given the planet-sized scale of the climate crisis and the ever-dwindling carbon budget to contain global warming within 1.5–2°C,[80] all sectors, including hard-to-decarbonize heavy industries, must reduce their GHG emissions as much and as rapidly as possible, and long-lived[81] CO_2 needs to be removed from the atmosphere. Doing both requires the massive deployment of all of the carbon sequestration pathways discussed in this chapter, with their potential prioritized in the consideration of land, water, nutrient, and economic resource constraints.

Policy and financial support are paramount. All CO_2 sequestrations can benefit from a robust carbon-price environment, for instance. Even with technologies that still need major cost reductions to achieve global adoption, the overall investments required for scale-up is relatively affordable in light of economic losses brought on by worsening climate disasters and in comparison to the historical funding of the fossil-fuel industries. Setting up a vibrant carbon sequestration industry is certainly a challenge, but one for which we already have technical and financial means.

STARTUP CASE STUDY
CarbonCure Technologies

by Moon K. Kim

CarbonCure is a Canadian, Nova Scotia-based startup that is one of the most commercially mature in the field of concrete curing, with ongoing projects in the Asia-Pacific region. It sources for captured CO_2 from industrial emitters to inject a precise dosage into a concrete mix. The CO_2 then reacts with calcium ions of the cement to form calcium carbonate, a solid mineral that gets embedded in the concrete permanently. This mineralization increases the compressive strength of the concrete by up to 10% while not affecting any other fresh or hardened properties of the stuff, such as finishing, color, texture, or durability. The method enables the reduction of cement content by 7% under the same strength requirements, thus lightening the carbon footprint of the resulting concrete products. This translates to 14.8 kg of CO_2 saved per cubic meter of ready-mix concrete and 14.8–23.7 kg/m3 for precast concrete.

The startup's technology retrofits into conventional concrete operations/plants with two pieces of equipment. The first is a Valve Box that connects to an onsite liquid CO_2 tank to automatically deliver a metered quantity of CO_2 into the concrete mix based on its cement content (see Figure 9.10). The second is the Control Box, which syncs with the plant's

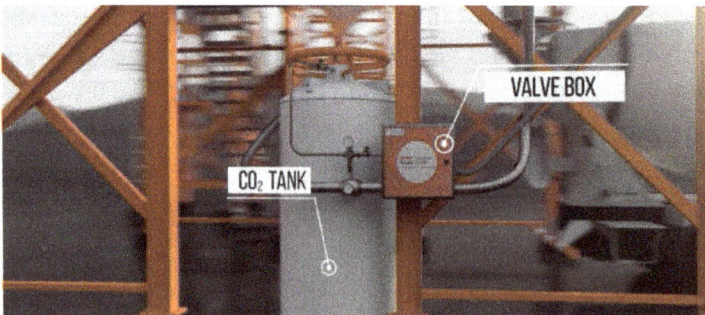

Figure 9.10. CarbonCure's Valve Box setup.

software to seamlessly integrate with existing batching or loading processes and monitors the performance of the Valve Box in real-time. The equipment installation is completed within hours in one visit, and ongoing customer support is provided remotely using telemetry gathered from the Control Box.

CarbonCure licenses its technology to concrete producers for monthly fees, requiring no upfront capital investment, and the technology costs for producers are offset by their cement savings.

In January 2018, the startup demonstrated the world's first integrated carbon capture and utilization from cement for concrete production in partnership with three companies: Argos USA (a subsidiary of Colombian cement company Argos SA and owner of cement plants and concrete operations in the US), Sustainable Energy Solutions (SES) (a Utah-based cryogenic carbon capture technology provider now owned by Chart Industries), and Praxair, an industrial gases company. CO_2 emissions from Argos' cement plant in Alabama were captured by SES, transported by Praxair, and then injected into concrete produced by Argos in Atlanta using CarbonCure's technology.

Building materials and construction generate 11% of global GHG emissions, and the world's building stock is expected to double by 2060, much of it in Asia. Given the scale and urgency to reduce embodied carbon in new buildings, it is not surprising that CarbonCure attracted interest from Asia-based concrete producers. The startup signed a strategic partnership agreement with Singapore-based Pan-United Corporation in November 2018, less than a year from its first pilot demonstration.

Pan-United is Singapore's largest ready-mix concrete and cement supplier, with overseas operations in Malaysia, Indonesia, and Vietnam. It is also the first concrete company to receive "Leader" certification from the Singapore Green Building Council, formally introducing the two companies. Under the strategic partnership, Pan-United is to install CarbonCure's technology at its concrete batching plants in Singapore, with a potential to save over 4 ktpa of CO_2 at each plant. CarbonCure's concrete has been used in various structures in Singapore, including commercial and residential buildings and infrastructure projects.

Whilst the startup's valuation is not public, and there are only estimations on the total funding received to date (potentially over

USD 100 million in venture capital financing), the cap-table includes the likes of Bill Gates' Breakthrough Energy Ventures and Amazon's Climate Pledge Fund. In early 2021, CarbonCure raised additional equity funding from venture capital firms and corporate investors, including Mitsubishi Corporation, Japan's largest trading company operating ten businesses across the globe, two of which are involved in real estate and infrastructure. The startup intends to use the capital to accelerate commercialization beyond the 300 concrete plants already using its technology and spur international expansion timelines, particularly in Europe and Asia. It is on a mission to remove 500 Mtpa of CO_2 by 2030, equivalent to taking more than 100 million cars off the road.

CORPORATE CASE STUDY
Sinopec

by Moon K. Kim

China is home to the largest and some of the youngest assets for coal-fired power plants and cement, iron and steel, and chemical plants. CC(U)S may be the only commercially available solution in the near future that can help these industries with the low-carbon transition. The country's potential capacity for geological carbon storage is estimated at a total of 425 Gt, equivalent to 40 years of its current emissions. Fortunately, 45% of emissions from these energy-intensive industries in China are within 50 km of potential storage sites, and 65% within 100 km.

EOR technology was first developed in China in the 1960s, and field tests have shown about a 10% increase in oil recovery. The theoretical carbon storage capacity of depleted onshore oil reservoirs is estimated at 3.8 Gt, with most of the active oilfields facing depletion after decades of production. However, extensive application of EOR may not be possible in China because the geophysical structure of most of these reservoirs shows low permeability and many fault planes.

Beijing-headquartered China Petrochemical Corporation (中国石油化工集团公司, Sinopec Group) is the world's largest energy company by revenue and the largest state-owned enterprise in China. The company is vertically integrated, from exploration and production to refining, marketing and distribution, and chemicals. Sinopec is also Asia's most advanced player in CCUS in terms of commercial projects in operation and development.

The Zhongyuan Oilfield Refinery facility (see Figure 9.11), commissioned in 2015 with a total capital investment of RMB 150 million, captures 0.1 Mtpa of CO_2 from the flue gas of a fluid catalytic cracking unit. Using a proprietary amine-based solvent chemical absorption process developed by Sinopec, this capture plant has achieved savings of 18% in energy consumption and 23% in solvent usage compared to typical lower concentration capture plants. The overall project generates an annual savings of RMB 39 million (USD 6.12 million) in the avoided purchase cost of CO_2 at RMB 390 (USD 61.2) per tonne.

Figure 9.11.   Photo of a carbon capture plant at Sinopec's Zhongyuan Oilfield Refinery.

The Qilu Petrochemical Refinery facility (in development) is designed to capture 0.4 Mtpa of CO_2 from a coal/coke gasification-syngas plant and transport the captured gas by pipeline over 75 km to the Shengli onshore oilfield for utilization in EOR.

In construction and scheduled to be commissioned by the end of 2021 at the time of writing, the Shengli facility will capture 1 Mtpa of CO_2 from the flue gas of a large coal-fired power plant in Shandong's Dongying City that powers the Shengli Oilfield. The power plant has been retrofitted with Sinopec's proprietary amine-based post-combustion capture technology. The captured CO_2 will be transported by pipeline over 80 km for EOR at the Shengli Oilfield.

The company currently has 36 EOR projects in China, including pre-commercial pilot and demonstration facilities, and it captured around 1.3 megatonnes of CO_2 in 2020 — the most recent year with accurate data. Sinopec is accelerating the implementation of CCUS projects as part of its low-carbon transition and plans to build more megatonne-scale CCUS facilities across its oil and gas fields in the next five years. This is a welcome start for a company that emitted 171 megatonnes of CO_2 in 2020 and is aligned with China's pledge to reach peak emissions by 2030 and the company's own target for carbon neutrality by 2050.

INTERVIEW
Dr. Anastasia Volkova
Co-Founder and
Chief Executive Officer, Regrow

by Rachel Fleishman

Regrow is one of the young-but-pioneering companies advancing MRV for ecosystem credits for regenerative agriculture. Focused on the mission to "power the business case for sustainable agriculture", Regrow uses scientifically vetted crop and soil models, connection to farm management platforms, and satellite imagery to create a fully independent MRV system.

Ukrainian Anastasia Volkova has a Ph.D. in Aerospace, Aeronautical, and Astronautical Engineering but deserves an honorary degree in high-impact juggling. She finished raising USD 38 million in Series B financing, which include Galvanize Climate Solutions, Time Ventures, and Rethink Impact while helping her mother escape the war in Ukraine. Dr. Volkova took the time to step back and reflect on her personal history, Regrow's trajectory, and the power of regenerative agriculture.

Give us your elevator pitch: What does Regrow do?

Regrow offers solutions that improve the environmental impact of the agriculture industry by providing measurement, reporting, and verification tools to farmers and food companies. We provide the computational backbone for implementing regenerative agriculture programs that are tailored to the needs of the farmer's fields. The result is products that end up on supermarket shelves that have a lower impact.

Consumers benefit in two ways. They get a healthier product and a sense of allegiance to the greater brand purpose: changing the food system for the better. In short, a simple purchase decision gives them a part in the solution to a pressing global problem — climate change.

Farmers get rewarded in several ways. They can be paid for adopting climate-smart practices, generating carbon credits, or selling low-emissions products with on-pack labeling. At the same time, farmers can gradually cut the cost of inputs (fertilizers and chemicals) without productivity loss — or even with productivity gain — as soil health improves in the long term. Finally, there is increased investor interest in farming enterprises that show environmental sustainability and improved resiliency.

For Regrow, the bottom line is impact. We help organizations that are passionate about agriculture transform themselves from being part of the climate problem to becoming part of the solution.

What inspired you to start the company?

When I was getting my degree in aeronautical engineering, it dawned on me that I was about to embark on a career in an industry that is contributing to the most important problem of our times — climate change. I could see how there might be ways that I could help this industry improve, but the corporate path seemed lengthy and uncertain — there was no clear path to make an impact. I was seeking more immediate ways to get my hands on the climate change problem and directly apply my skills.

The issues caused by climate change are urgent. Our generation cannot be patient; we need to act! I took several positions in startups, where I learned what it meant to start and run a startup. I started looking for ways that I could apply my knowledge of remote sensing and software development to solve real-world problems. I discovered the challenges faced by the agriculture industry, which was a lack of data to make informed decisions or a lack of incentives. This is where I stayed.

Today, Regrow is helping several consumer-packaged goods companies reduce their product emissions. In practice, this means working with farmers on adopting regenerative agriculture practices, including no-till, cover cropping, more targeted use of fertilizer, reduction of emissions from on-farm equipment use, and the adoption of smart irrigation techniques that require less water and energy.

Our software platform enables farmers to discover, implement, and get rewarded for farming methods that are more environmentally friendly. By assisting farmers to adopt regenerative agriculture and measuring the positive impact they are making, we also help food companies reward them. We are uniquely positioned to help brands not only baseline the environmental impact of their agricultural supply chain but also to deliver impact and measure the change in a manner that is visible to the consumer.

What is most gratifying for me is finding and addressing the most pressing problems for the industry and developing technology and business models that can help the industry rally behind a common cause, such as fighting climate change and delivering healthier, more nutritious food.

What presents the biggest barriers to adoption and scaling today?

Modeling real-world processes in agriculture is quite a data-hungry process. Given that agriculture is one of the least digitized industries, we often face the challenge of missing data to be able to scale some of our solutions faster. It's not possible to acquire all needed data. High-quality ground truth data from farms, business working with farms, or government agencies is necessary to evaluate the quality of our models. There are regions where such data is abundant, but there are many regions where it is still scarce, and we are partnering with many organizations to create a business case for collecting more good data.

Where are you deployed now and how do you see that expanding in the future, especially in Asia-Pacific?

Our Agronomy solution is available globally and has been used on farms in 45 countries. Our Sustainability and Carbon solutions, including the MRV platform, are currently available in North America and are being prepared for roll-out in Latin America, Australia, and Europe. We have Agronomy deployed in Asia and collaborate with several not-for-profit organizations in advancing regenerative agriculture practices in the region. On the commercial side, we are helping a few Asian companies discover the business opportunities of investing in climate-smart agricultural practices. For example, in Vietnam, we help quantify the benefits of the adoption of climate-smart farming practices by rice farmers and help popularize the program among locals by creating decision support tools that are accessible and easy to use. I hope this will grow into a transformation of the industry in the region under the leadership of these innovators.

What is the most exciting thing you're working on right now?

We are supporting large-scale, farmer-facing programs incentivizing practices that build soil health and increase soil carbon storage. I see big potential for these programs because they simultaneously help cool the planet, leverage consumer interest in climate-smart products, improve farmer livelihoods, and build soil health that will sustain food for future generations. My dream is to turn all of the world's arable acres green.

Endnotes

1. Cameron Hepburn, Ella Adlen, John Beddington, *et al.*, "The technological and economic prospects for utilization and removal", *Nature Communications* **575** (2019) 87–97.

2. Sequestration of methane (CH4), the other carbon-based primary greenhouse gas that is 84 times stronger in its global warming potential but present in much smaller quantity than CO2, is still in infancy and remains in early research domains.

3. Durable carbon means a form of carbon in the technosphere that is not easily emitted into the atmosphere even at the end of the lifecycle of the industrial product in which it is embedded.

4. Cost of planting plus ongoing maintenance until the forests become self-sustaining.

5. USD 21/ha/year average cost in China (See Ying Wang, Qi Zhang, Richard Bilsborrow, *et al.*, "Effects of payments for ecosystem services programs in China on rural household labor allocation and land use", *Land Use Policy* **99** (2020) 105024) and USD 440/ha average life-time cost in several African and Latin American countries (See Heng Ding, Sofia Faruqi, Andrew Wu, *et al.*, "Roots of prosperity: The economics and finance of restoring land", World Resources Institute, 19 December 2017); annual sequestration of 5 tonnes of CO2 by one acre of young forest assumed for 30 years of growth.

6. Formed by burning biomass at high temperatures in the absence of oxygen and added to acidic soils to increase agricultural productivity.

7. Ben Elgin, "A top US seller of carbon offsets starts investigating its own projects", *Bloomberg*, 5 April 2021.

8. Equinor today.

9. 1 megatonne = 1 million tonnes

10. Geological formation of sedimentary rocks containing salt water.

11. Such as in natural gas processing and production of fertilizer, chemicals, and hydrogen.

12. Combustion exhaust gas exiting to the atmosphere through a pipe or channel.

13. The world's largest CC(U)S facility is ExxonMobil's Shute Creek Gas Processing Plant in Wyoming (commissioned in 1986), capturing 7 Mtpa of CO2 from natural gas processing.

14. International Energy Agency, "Special report on carbon capture utilization and storage", 2020, p. 27.

15. Ibid, p. 114.

16. Such as porosity, permeability, and geometry of rock layers.
17. Fractured surfaces between geological blocks through which underground gases can escape.
18. Estimated at 6,000–42,000 Gt across onshore sites (mostly in deep saline aquifers and depleted oil and gas fields) and 2,000–13,000 Gt across offshore sites; saline aquifers have been considered ideal sites for CCS due to their relative abundance and vast storage capacity (i.e., several thousands of Gt versus hundreds of Gt in oil and gas fields), but little is known about their geophysical structure, and limited exploration may increase risk of leakage.
19. 1 gigatonne = 1 billion tonnes; 1 tonne = 1,000 kg
20. International Energy Agency, "Special report on carbon capture utilization and storage", p. 114.
21. Global CCS Institute, "Global status of CCS", 2020.
22. Including manufacturing of synthetic fuels, chemicals, plastics, food, and beverage.
23. International Energy Agency, "Special report on carbon capture utilization and storage", p. 117.
24. Around USD 15–30/tonne; International Energy Agency, "Special report on carbon capture utilization and storage", p. 117.
25. The world's largest operational CCUS installation on a power plant with 1.4 Mtpa capture capacity, which is equivalent to capturing emissions from more than 300,000 passenger vehicles. This estimate is based on a typical passenger vehicle emitting 4.6 tonnes of CO_2 every year; United States Environmental Protection Agency, "Greenhouse gas emissions from a typical passenger vehicle", 2021.
26. According to NRG (the facility's operator), crude oil prices higher than USD 60–65/barrel are required to cover their operating costs.
27. Global CCS Institute, "Global status of CCS".
28. Ibid.
29. Ibid.
30. Nineteen out of the 37 new facilities.
31. The Carbon Storage Assurance Facility Enterprise initiative is developing geological storage hubs, each capable of storing >50 Mt CO_2, to reduce the unit cost of storage and investment risk for CC(U)S clusters.
32. Malaysia is prioritizing CCS development in high-CO_2 gas fields, with its state-owned energy company Petronas planning a 3.7 Mtpa offshore project near Sarawak (to be commissioned by the end of 2025) to reduce carbon emission from flaring; Petronas has recently partnered with ExxonMobil to

jointly explore potential large-scale CCS projects in offshore Peninsular Malaysia.

33. Two in Norway and one each in Australia, Canada, Qatar, and the US.
34. Global CCS Institute, "Global status of CCS".
35. Ibid.
36. Global CCS Institute, "Global status of CCS".
37. Oil and Gas Climate Initiative, "CCUS in China", 2021.
38. Global CCS Institute, "Global status of CCS".
39. Oil and Gas Climate Initiative, "CCUS in China".
40. Partners include Add Energy Group, Commonwealth Scientific and Industrial Research Organization, Mitsui O.S.K. Lines, Kyushu Electric Power, Osaka Gas Co, TechnipFMC, and Tokyo Gas Australia.
41. Notable member companies include Inpex, JGC, Mitsubishi, Mitsui O.S.K. Lines, MUFG Bank, Nippon Steel, PT Pertamina, Sumitomo, and Tokyo Gas.
42. 1 kilotonne = 1,000 tonnes.
43. International Energy Agency, "Special report on carbon capture utilization and storage", p. 81.
44. Christopher Consoli, "Bioenergy and carbon capture and storage", Global CCS Institute, 14 March 2019.
45. Sabine Fuss, William F. Lamb, Max W. Callaghan, *et al.*, "Negative emissions — Part 2: Costs, potentials and side effects", *Environmental Research Letters* **13** (2018) 063002.
46. If a waste-to-energy plant can capture and permanently store higher quantities of biogenic CO_2 (from biomass waste-based-fuels such as paper, cardboard, wood, food waste, and garden trimmings) than CO_2 produced from combustion of fossil-fuel based waste such as plastics, then the plant's overall emissions become negative.
47. Consoli, "Bioenergy and carbon capture and storage"; the upper range of the 1.5°C warming scenarios is 16 Gtpa of CO_2 removal by 2100.
48. Mathilde Fajardy and Niall MacDowell, "Can BECCS deliver sustainable and resource efficient negative emissions?", *Energy and Environmental Science* **10** (2017) 1389–1426.
49. Available in two standardized capture capacities of 40 ktpa and 100 ktpa of CO_2.
50. International Energy Agency, "Special report on carbon capture utilization and storage", p. 101.
51. Caused by chemical breakdown of the amine molecules due to reactions with other gases and pollutants as well as thermal deterioration.

52. The company raised additional USD 150 million in May 2022 from Chevron, CEMEX, Marubeni, Saudi Aramco, among others, to deploy fully modular carbon capture units "CycloneCC" with ten times a smaller footprint (launched in 2021).
53. CO2 binding to the surface of a solid adsorbent.
54. The nano-engineered adsorbents catch and release CO2 rapidly (i.e., less than 60 seconds versus hours with other technologies), increasing through-put, reducing physical footprint, and thus cutting down the capital cost of the capture system.
55. It should be noted that these are not durable carbons.
56. Such as humus, decaying and fecal material, microbes, fungi, or soil invertebrates.
57. Higher clay content increases soil carbon by physically protecting organic matter from decomposition.
58. R. Lal, "Soil carbon sequestration impacts on global climate change and food security", *Science* **304** (2004) 1623–1627.
59. Jose Navarro-Pedreño, María Belén Almendro-Candel, and Antonis A. Zorpas, "The increase of soil organic matter reduces global warming, myth or reality?", *Science* **3** (2021) 18.
60. Use of synthetic fertilizer produced via the 100-year-old Haber–Bosch process contributes to 7% of global GHG emissions, especially in runoff of nitrous oxide, which has 265 times the global warming potential of CO2.
61. Climate Action Reserve, "Soil enrichment protocol", 2020.
62. 1 hectare = 10,000 square meters
63. Jeremy Proville, Robert Parkhurst, Steven Koller, *et al.*, "Agricultural offset potential in the United States, economic and geospatial insights", The Environmental Defense Fund, 13 November 2020.
64. Such as holistic planned grazing without irrigation or use of fertilizers and chemicals.
65. Principles include no tillage, fertilizer, pesticides or herbicides, weeding, and pruning.
66. Bettina and Tara, "Empowering smallholder farmers: How Asian countries are building up their agriculture backbones", *Agrifood*, 19 February 2021.
67. Direct air carbon capture and storage.
68. DAC uses water in hot and dry conditions; Katie Lebling, "Direct air capture: Resource considerations and costs for carbon removal", World Resources Institute, 29 January 2021.

69. Approximately 400 parts per million, which is 1% of the CO2 concentration in the flue gas of a gas-fired power station; Global CCS Institute, "Global status of CCS".

70. DAC has not reached large-scale demonstration, with capture cost estimates varying widely from USD 135/tonne of CO2 to USD 345/tonne (see Figure 9.7) that are up to 23 times higher than BECCS capture costs of USD 15–60/tonne; the breakeven cost might be USD 100/tonne as large customers generally pay USD 65–110/tonne for CO2 used in commercial and industrial processes.

71. International Energy Agency, "Special report on carbon capture utilization and storage", p. 84.

72. A 50 tpa capacity started in 2017 and developed in partnership with Iceland's geological carbon storage developer, CarbFix, and geothermal energy company, On Power.

73. Lebling, "Direct air capture: Resource considerations and costs for carbon removal".

74. Company website.

75. The first 500 ktpa to come online by 2026.

76. Assuming large-scale deployment of 1 Mtpa facilities.

77. Klaus S. Lackner and Habib Azarabadi, "Buying down the cost of direct air capture", *Chemistry Research* **60** (2021) 8196–8208.

78. James Temple, "What it will take to achieve affordable carbon removal", *MIT Technology Review*, 25 June 2021.

79. Production of cement, a binder used in making concrete, is the second-largest industrial emitter of CO2, responsible for 8% of global GHG emissions; International Energy Agency, "Cement technology roadmap plots path to cutting CO2 emissions 24% by 2050", 2021.

80. Temperature rises beyond 2°C will reinforce warming and amplifying feedback processes of the Earth's carbon cycle, making climate change mitigation and adaptation efforts increasingly more difficult and ultimately leading to irreversible effects that will be catastrophic to the modern world.

81. Once released, CO2 stays in the atmosphere for 300 to 1,000 years.

EPILOGUE

by Tony Á. Verb

This book has been the result of a year of research, writing, and fine-tuning by more than a dozen contributors, all of whom brought their wealth of experience and insights into this project. Even with all that, we are still only scratching the surface to highlight the breadth and scale of interventions necessary to keep temperatures from rising above 1.5°C. We know too well that each chapter of this book would require a full book's length of contemplation and deep diving in order to cover the topic with an analyst's desire for depth and detail.

Nonetheless, we thought this work was important, because there is far too little space dedicated for Asia-Pacific in the global climate innovation and investment discourse. The book's aim was to provide a snapshot of the key sectors and some important players, along with notable examples of companies, organizations, and individuals in the region.

Curating these players and examples was harder than we first thought, simply because so much is happening. Selecting the "leading" corporation, the "best" startup, and the "most impressive" leader to feature was an impossible task as there were so many great ones to choose from. We decided early on not to aim for a list of the "best", but rather a diverse list that reflects the realities, breadth of the decarbonization-related innovation, and investment activity in the region. We included indigenous as well as foreign firms and individuals, some niche players, and some household names, making sure that they contributed to the scope of the book on merit.

We learned a lot together, and we processed and presented a great amount of data, but we deliberately did not try to answer complex and challenging questions about prioritization and hard choices that have to be made between different technologies and policies. For instance, we now know that cca. 690 million tonnes of food is lost and wasted in Asia — for the most part, immediately after harvest. That is a significant societal and carbon issue, but how do we prioritize that to electricity generation, which is, directly and indirectly, responsible for more than 50% of today's CO_2 emissions?

And what about the industry sector's 30% contribution of all greenhouse gas (GHG) emissions and the tremendous waste issue it creates? Or is transportation a higher priority to decarbonize because it is second only to power generation in its emissions share? As we read in the transport chapter, there is a pressing need to disrupt all forms of transport. How do we do that? At the turn of the last century, there was a massive change in mobility as well. Henry Ford once remarked: "People didn't want faster horses. They wanted less horseshit". What does that mean to us and policymakers in general?

Most of our Contributors wrote in their recommendation and call-to-action statements that more regulation, commitment to targets, openness to innovation, investment, and collaboration are desired. Our focus remained on great people, insightful case studies, and anecdotes. Our only regret is that we could not include more of all that.

There is plenty of good news to be excited about: when we started this project over the summer of 2021, the tides were already changing in awareness, attitude, and commitment to climate-related innovation and investment. We did not know that whilst we were writing the book, gears would be shifted so dramatically. That had a lot to do with COP26, as expectations rose beforehand because of the media, NGOs, and global popular opinion.

Between June and November 2021, when the UN's Climate Conference took place in Glasgow, India, Saudi Arabia, Thailand, the United Arab Emirates, and Vietnam joined the list of countries with net-zero targets. A couple of months later, Singapore came out with a detailed plan and milestones in introducing a carbon tax and other measures to decarbonize its economy. Adding these countries to China, Hong Kong SAR, Japan, and South Korea, economies that made their net-zero commitments earlier in 2020 and 2021, signifies a change whose significance cannot be overstated. These net-zero commitments add up to more than 40% of the world's CO2 emissions. Without them, no global target is achievable, and no effort makes much sense. With them, we have a realistic shot at our collective climate goals.

This is only the start, however. Now, it is time for action and action beyond what we observed and introduced in the book. Regulations must change quickly, capital must flow aplenty, and decarbonizing technologies must scale across borders and sectors. Whilst we decided to write this book with a positive and optimistic mindset, we never claimed that all is rosy. More needs to be done as the actual baseline is much worse than we thought when we first started this book. In the meantime, several IPCC reports have been published, ushering in more urgency and shutting the door on any further debate on whether climate change is happening and whether it is anthropogenic.

At the same time, our optimism is reinforced by the same reports and our research. Firstly, IPCC confirmed that it is not too late to act, and if we focus our efforts and resources, we can keep global warming under the 1.5°C target. Secondly, action is already underway. We saw this across the region and across industries. The featured case studies and more than 100 examples that were mentioned across the book only reveal the tip of the iceberg. We had no space to publish more due to our sensible length cap from our publishing partner.

Venture capital funds have been launched and are about to be announced, government and institutional funding is being chanelled to green and carbon-related investments, new startups are founded on a daily basis, and the world's best innovators are expanding in the region. What we expect over the coming years is an explosion of decarbonization-specific investments in technology, infrastructure, and human capital. We believe that this is the greatest opportunity of the decade for the public sector to make a difference in industrial and societal development and for the private sector to contribute to the process without compromising its commercial interests.

The authors, editors, and contributors of this book, along with all the featured companies, organizations, and individuals, will certainly continue to play a significant role in the region. Any reader is welcome to connect. As much as decarbonization is a shared responsibility, it can be a shared opportunity too. Let us work together towards it.

Everyone is welcome to reach out via: hello@carbonless.xyz

Index

www.ingramcontent.com/pod-product-compliance
Lightning Source LLC
Chambersburg PA
CBHW052118230326
41598CB00080B/3809